Students of culture have been increasingly concerned with the ways in which cultural values are "inscribed" on the body. These essays go beyond this passive construal of the body to a position in which embodiment is understood as the existential condition of cultural life. From this standpoint embodiment is reducible neither to representations of the body, to the body as an objectification of power, to the body as a physical entity or biological organism, nor to the body as an inalienable center of individual consciousness. This more sensate and dynamic view is applied by the contributors to a variety of topics, including the expression of emotion, the experience of pain, ritual healing, dietary customs, and political violence. Their purpose is to contribute to a phenomenological theory of culture and self – an anthropology that is not merely about the body, but from the body.

Embodiment and experience

Cambridge Studies in Medical Anthropology 2

Editors

Ronald Frankenberg, *Centre for Medical Social Anthropology, University of Keele*

Byron Good, *Department of Social Medicine, Harvard Medical School*

Alan Harwood, *Department of Anthropology, University of Massachusetts, Boston*

Gilbert Lewis, *Department of Social Anthropology, University of Cambridge*

Roland Littlewood, *Department of Anthropology, University College London*

Margaret Lock, *Department of Humanities and Social Studies in Medicine, McGill University*

Nancy Scheper-Hughes, *Department of Anthropology, University of California, Berkeley*

Medical anthropology is the fastest growing specialist area within anthropology, both in North America and in Europe. Beginning as an applied field serving public health specialists, medical anthropology now provides a significant forum for many of the most urgent debates in anthropology and the humanities.

Medical anthropology includes the study of medical institutions and health care in a variety of rich and poor societies, the investigation of the cultural construction of illness, and the analysis of ideas about the body, birth, maturation, ageing, and death.

This new series includes theoretically innovative monographs, state-of-the-art collections of essays on current issues, and short books introducing the main themes in the subdiscipline.

1 Lynn M. Morgan, *Community participation in health: the politics of primary care in Costa Rica*

Embodiment and experience

The existential ground of culture and self

Edited by

Thomas J. Csordas
Case Western Reserve University

CAMBRIDGE
UNIVERSITY PRESS

Published by the Press Syndicate of the University of Cambridge
The Pitt Building, Trumpington Street, Cambridge CB2 1RP
40 West 20th Street, New York, NY 10011–4211, USA
10 Stamford Road, Oakleigh, Melbourne 3166, Australia

First published 1994

Printed in Great Britain at the University Press, Cambridge

A catalogue record for this book is available from the British Library

Library of Congress cataloguing in publication data

Embodiment and experience: the existential ground of culture and self
/ edited by Thomas. J. Csordas.
    p.   cm. – (Cambridge studies in medical anthropology; 2)
  ISBN 0 521 45256 2 (hardback). – ISBN 0 521 45890 0 (paperback)
  1. Body, Human–Social aspects. 2. Body, Human–Symbolic aspects.
3. Medical anthropology. I. Csordas, Thomas J. II. Series
GN298.E43 1994
306.4'61–dc20   93–45993   CIP

ISBN 0 521 45256 2 hardback
ISBN 0 521 45890 0 paperback

CE

Contents

Illustrations

Contributors

JACK M. BARBALET Department of Sociology, Australian National University

ANNE E. BECKER Department of Psychiatry, Massachusetts General Hospital, Harvard Medical School

THOMAS J. CSORDAS Department of Anthropology, Case Western Reserve University

E. VALENTINE DANIEL Department of Anthropology, University of Michigan

LINDSAY FRENCH Department of Anthropology, Harvard University

JEAN JACKSON Anthropology/Archaeology Program, Massachusetts Institute of Technology

JANIS H. JENKINS Departments of Anthropology and Psychiatry, Case Western Reserve University

CAROL LADERMAN Department of Anthropology, City College, City University of New York

SETHA M. LOW Program in Psychology: Environmental Psychology, Graduate Center, City University of New York

MARGOT L. LYON Department of Archaeology and Anthropology, Australian National University

THOMAS OTS Fachbereich Sozialwesen, Hochschule für Technik, Wirtschaft und Socialwesen Zittau/Görlitz, Germany

TERENCE TURNER Department of Anthropology, University of Chicago

MARTHA VALIENTE Department of Psychiatry, The Cambridge Hospital, Harvard Medical School

KATE WININGER Department of Philosophy, University of Southern Maine

CATHY WINKLER Department of Sociology and Anthropology, Tuskegee University

Preface

It is probably no fluke of intellectual history that a turn toward the body in contemporary scholarship in the human sciences has coincided with the realization that the postmodern condition is now the uneasy condition of all intellectual activity. If behind the turn to the body lay the implicit hope that it would be the stable center in a world of decentered meanings, it has only led to the discovery that the essential characteristic of embodiment is existential indeterminacy. Each in their own way, the contributors to this volume explore this indeterminacy, in which embodiment is reducible neither to representations of the body, to the body as an objectification of power, to the body as a physical entity or biological organism, nor to the body as an inalienable center of individual consciousness.

For anthropologists the culminating event in the turn to the body was the 1990 annual meeting of the American Ethnological Association dedicated to the theme of "The Body in Society and Culture." The majority of papers included in the present volume were originally presented in sessions at that meeting. The core of the volume originated in a session on "The Body as Existential Ground of Culture," organized by me and Susan DiGiacomo. Most of the others were recruited from sessions in which individual participants were clearly struggling with the problem of how serious consideration of embodiment implied a change in our conceptualization of culture, self, and experience. How this change will be realized remains to be felt. For the present, we offer these original essays as a step toward an anthropology that is not merely about the body, but from the body.

Thomas J. Csordas

Introduction: the body as representation and being-in-the-world

Thomas J. Csordas

Much has been written about the body in recent years. Beginning in the early 1970s, and with increased energy in the late 1980s, the body has assumed a lively presence on the anthropological scene, and on the stage of interdisciplinary cultural studies. Feminist theory, literary criticism, history, comparative religion, philosophy, sociology, and psychology are all implicated in the move toward the body. Anthropologists with interests ranging across medical and psychological anthropology, the anthropology of space, material culture, practice theory, performance theory, critical theory, and even cognitive anthropology have problematized the body in recent writings.

In her keynote address to the 1990 annual meeting of the American Ethnological Association dedicated to the theme of "The Body in Society and Culture," Emily Martin suggested that although the widespread interest in the body may be accounted for by the contemporary centrality of the body in Western social forms, it may also be due to the contemporary historical moment in which "we are undergoing fundamental changes in how our bodies are organized and experienced" (1992: 121). Citing Lévi-Strauss's observation that academic attention seems to become focused on phenomena precisely when they are ending, she suggests that we are seeing "the end of one kind of body and the beginning of another kind of body" (ibid.: 121).

Recent scholarship in the social sciences and humanities would appear to support Martin's claim. The kind of body to which we have been accustomed in scholarly and popular thought alike is typically assumed to be a fixed, material entity subject to the empirical rules of biological science, existing prior to the mutability and flux of cultural change and diversity and characterized by unchangeable inner necessities. The new body that has begun to be identified can no longer be considered as a brute fact of nature. In the wake of Foucault (e.g. 1979, 1980), a chorus of critical statements has arisen to the effect that the body is "an entirely problematic notion" (Vernant 1989: 20), that "the body has a history" in that it behaves in new ways at particular historical moments (Bynum 1989:

1

171), and that the body should be understood not as a constant amidst flux but as an epitome of that flux (A. Frank 1991: 40).

Others have argued that, due to the destabilizing impact of social processes of commodification, fragmentation, and the semiotic barrage of images of body parts (Kroker and Kroker 1987: 20), the human body can no longer be considered a "bounded entity." In the milieu of "late capitalism" and "consumer culture," with its multiplicity of images that stimulate needs and desires and the corresponding changes in material arrangements of social space, the body/self has become primarily a performing self of appearance, display, and impression management (Featherstone 1991: 187, 192). Fixed "life cycle" categories have become blurred into a more fluid "life course" in which one's look and feel may conflict with one's biological and chronological age (Featherstone and Hepworth 1991). The goals of bodily self-care have historically changed from spiritual salvation to enhanced health, and finally to a marketable self (ibid.: 170; cf. Foucault 1986 and Bordo 1990: 85). The dieter's techniques are not directed primarily toward weight loss, but toward the formation of body boundaries to protect against the eruption of the "bulge," and they serve the purposes of social mobility more than the affirmation of social position (Bordo 1990: 90, 95). The asceticism of inner body discipline is no longer incompatible with outer body hedonism, but has become a means toward it; one not only exercises to look good, but wants to look good while exercising (ibid.: 171, 182), as can be attested to by anyone observing that in some health clubs women apply makeup *before* their workout. It stands in sharp contrast not only to early historical periods, but to other societies such as that of Fiji in which the cultivation of bodies is not intended as an enhancement of a performing self, but is regarded as a responsibility toward the community (Becker, Chapter 4 of this volume).

Donna Haraway argues forcefully that "Neither our personal bodies nor our social bodies may be seen as natural, in the sense of existing outside the self-creating process called human labour" (1991: 10), and that as a feature of ideology "the universalized natural body is the gold standard of hegemonic social discourse" (Haraway 1990: 146). According to Haraway, the appropriate alternative to the naturalized and essentialized body is not relativism, which is only the inverse of the totalizing perspective, a view which denies embodiment by "being nowhere while claiming to be everywhere equally" (1991: 191). Instead of relativism she advocates the recognition of *location*, that is, non-equivalent positions in a substantive web of connections. The emphasis on location accepts the interpretive consequences of being grounded in a particular embodied standpoint – the consequences of relatedness, partial grasp of any situation, and imperfect communication. Theoretically, this situatedness extends to the domain of biology itself, as is evident in recent feminist theory that eliminates "passivity" as an intrinsic characteristic of the female body, and reworks both the distinction between sex and gender (Haraway 1991: 197–8), and the decoupling of female sexual pleasure from the act of conception (Jacobus 1990: 11, 22, 26; Bordo 1990: 103; Haraway 1990, Doane 1990; Keller 1990). With biology no longer a monolithic objectivity, the body is transformed from object to agent (Haraway 1991: 198; see also A. Frank 1991: 48). The body as an experiencing agent is evident in recent social science work on the experience of illness (Devisch and Gailly 1985; Kleinman 1988; Murphy 1987; Lock and Dunk 1987; Gordon 1990; Pandolfi 1991; Ots 1991; Kirmayer 1989, 1992; Good 1994), body image (G. Frank 1986), pain (Good et al. 1992), religious healing (Csordas 1990, 1993, 1994; Roseman 1991; Desjarlais 1992), and ethnographic practice itself (Scheper-Hughes and Lock 1987; Jackson 1989; Stoller 1989), as well as in the chapters included in the present volume.

The contemporary cultural transformation of the body can be conceived not only in terms of consumer culture and biological essentialism, but also in discerning an ambiguity in the boundaries of corporeality itself. Haraway points to the boundaries between animal and human, between animal/human and machine, and between the physical and non-physical (1991: 151–4). Feher, in his introduction to the influential *Fragments for a History of the Human Body*, places the boundary between human and animal or automaton (machine) at one end of a continuum whose opposite pole is defined by the boundary between human and deity (1989: 11). Examining what takes place at these cultural boundaries is critical, given the circumstances of corporeal flux and bodily transformation sketched above. With respect to religion, the question goes beyond the distinction between natural and supernatural bodies, or between natural corporeality and divine incorporeality, to the question posed by Feher of the kind of body that members of a culture endow themselves with in order to come into relation with the kind of deity they posit to themselves (1989: 13). If we are to assert that the body is a cultural phenomenon, religion is one domain of culture that offers evidence rich enough to help us grasp the significance of that assertion, and it is thus no coincidence that several of the chapters in the present volume take up the relation between religious experience and embodiment.

Another inescapable transformation of the body in the contemporary world is being wrought by the incredible proliferation of political violence of all types: ethnic violence, sexual violence, self-destructive violence, domestic violence, and gang violence. As much as any of the transformations sketched above, this one has to do with the very meaning of being human as being a body that can experience pain and self-alienation. From Scarry's (1985) examination of the dissolution of self in torture to Feldman's (1991) portrait of the denatured body that exists in the climate of permanent violence in Northern Ireland; from Scheper-Hughes's (1992) analysis of unarticulated bodily resistance to hegemonic oppression among impoverished residents of Brazilian slums, and again to the madness of "ethnic cleansing" and rape as a political weapon that characterizes the former Yugoslavia at the time of the writing of this introduction, the body appears as the threatened vehicle of human being and dignity. The moral and political urgency of this phenomenon is evident in the work of several contributors to the present volume.

Along with its critical and pragmatic implications for world civilization, the theoretical implications of the scholarly discovery that the body has a history and is as much a cultural phenomenon as it is a biological entity are potentially enormous. Also, if indeed the body is passing through a critical historical moment, this moment also offers a critical methodological opportunity to reformulate theories of culture, self, and experience, with the body at the center of analysis. The aims of this volume are to draw out some of those theoretical implications and to seize this methodological opportunity. Neither of these aims is to be taken for granted, since among anthropologists facing the "obsolescence of the body" and a related "death of the subject" the jury is still out as to whether the body will persist as a central analytic theme, the "existential ground of culture and self" (Csordas 1990), or whether interest in the body is merely an intellectual fad. At the 1990 meeting of the American Ethnological Society, dedicated to the theme of "the body in society and culture," it was evident that many participants were using the term "body" without much sense of "bodiliness" in their analyses, as if body were little more than a synonym for self or person. This tendency carries the dual dangers of dissipating the force of using the body as a methodological starting point, and of objectifying bodies as things devoid of intentionality and intersubjectivity. It thus misses the opportunity to add sentience and sensibility to our notions of self and person, and to insert an added dimension of materiality to our notions of culture and history.

What we are calling for here is a more radical role for the body than that typical in the "anthropology of the body" that has been with us since the 1970s. In studies that fall under that rubric, the body is an object or theme of analysis, often the source of symbols taken up in the discourse of cultural domains such as religion and social structure. Without attempting a bibliographical essay, I will summarize the approaches characteristic of the anthropology of the body in order then to distinguish a methodological standpoint more tailored to the above-stated aims.[1]

A premise of much of this literature is what we might call an "analytic body" that invites a discrete focus on perception, practice, parts, processes, or products. By perception I mean the cultural uses and conditioning of the five external senses plus proprioception (our sense of being in a body and oriented in space), as well what Kant (1978 [1800]) called the inner sense of intuition or sensibility. Practice includes everything that falls under Mauss's (1950) classic notion of techniques of the body – swimming, dancing, washing, ritual breathing in meditation, posture, the variations in batting stance among baseball players – in which the body is at once tool, agent, and object. Parts of our anatomy such as hair, face, genitals, limbs, or hands have long been of interest to anthropologists for the social and symbolic significance they bear. Bodily processes like breathing (not as a technique but, for example, as the sigh), blushing, menstruation, birth, sex, crying, and laughing are of interest in their cultural variation. Finally, a great deal of cultural meaning can be distilled from the treatment of body products such as blood, semen, sweat, tears, feces, urine, and saliva.

Other literature in this field concentrates on the "topical body," that is, an understanding of the body in relation to specific domains of cultural activity. The body and health, the body and political domination, the body and trauma, the body and religion, the body and gender, the body and self, the body and emotion, the body and technology are examples. The generation of abundant literatures on all these topical bodies has been quite recent and quite rapid, such that the body's existential ubiquity has become overwhelmingly apparent in scholarly production. This postmodern proliferation itself again begs the essentialist question of whether there is in fact any such thing as *the* body – whether the body is more than the sum of its topics. The paradoxical truth, in fact, appears to be that if there is an essential characteristic of embodiment, it is indeterminacy (Merleau-Ponty 1962; Csordas 1993).

Finally, there is what we might call the "multiple body," with the number of bodies dependent on how many of its aspects one cares to recognize. Mary Douglas (1973) called attention to the "two bodies," referring to the social and physical aspects of the body. Her distinction roughly reiterates that between mind and body, culture and biology. More precisely, Douglas differentiates between the use we make of our bodies and the way our bodies function, and emphasizes the way elements of physiology and anatomy can be taken up into the symbolic domain. Nancy Scheper-Hughes and Margaret Lock (1987) give us "three bodies," including the individual body, the social body, and the body politic. The first refers to the lived experience of the body as self, the second to representational uses of the body as a symbol of nature, society, and culture, and the third to the regulation and control of bodies. John O'Neill (1985) ups the ante to "five bodies." For O'Neill, the world's body refers to the human tendency to anthropomorphize the cosmos. The social body refers to the common analogy of social institutions to bodily organs and the use of bodily processes such as ingestion of food to define social categories. The body politic refers to models of city or country as the body writ large, forming the basis of phrases such as "head" of state or "members" of the body politic. The consumer body refers to the creation and commercialization of bodily needs such as for sex, cigarettes, labor-saving devices, or cars, a process in which doubt is created about the self in order to sell grace, spontaneity, vivaciousness, confidence, etc. The medical body refers to the process of medicalization in which an increasing number of body processes are subject to medical control and technology.

To greater or lesser degrees all these approaches study the *body* and its transformations while still taking *embodiment* for granted. In my view this distinction between the body as either empirical thing or analytic theme, and embodiment as the existential ground of culture and self is critical to capitalizing on the methodological opportunity identified above. But lest it be objected that if anything can be taken for granted it is embodiment, let us begin to reframe the problem this way. In his often-cited essay on "Techniques of the Body" Marcel Mauss (1950) argued that the body is at the same time the original tool with which humans shape their world, and the original substance out of which the human world is shaped. Yet of all the formal definitions of culture that have been proposed by anthropologists, none have taken seriously the idea that culture is grounded in the human body.[2] Why not then begin with the premise that the fact of our embodiment can be a valuable starting point for rethinking the nature of culture and our existential situation as cultural beings? I suggest that the promise of such a standpoint is to throw new light on questions traditionally asked by anthropologists and other scholars in the human sciences (see Fernandez 1990 for an example of a scholar reconsidering his own data in this way). It should also, as the chapters in this volume bear out, bring to light new questions and sources of data overlooked by thinkers in these fields. Finally, it offers the grounds for a fruitful rereading of the classic data of ethnography, where passages about bodily experience are tucked away in discussions of ritual and social organization, waiting to be rediscovered.

With regard to the last point, it is telling that what is perhaps the most vivid example of the body as a cultural phenomenon subject to cultural transformations is also one of the oldest in anthropology. Maurice Leenhardt, the anthropologist and missionary whose classic work on New Caledonian culture first appeared in 1947, described his discovery of the impact of Christianity on the cosmocentric world of the Canaques with the anecdote of a conversation between himself and an aged indigenous philosopher. Leenhardt suggested that the Europeans had introduced the notion of "spirit" to the indigenous way of thinking. His interlocutor contradicted him, pointed out that they had "always acted in accord with the spirit. What you've brought us is the body" (Leenhardt 1979 [1947]: 164). In brief, in

the indigenous world view the person was not individuated, but was diffu-
sed with other persons and things in a unitary sociomythic domain:

[The body] had no existence of its own, nor specific name to distinguish it. It was
only a support. But henceforth the circumscription of the physical being is completed, making possible its objectification. The idea of a human body becomes
explicit. This discovery leads forthwith to a discrimination between the body and
the mythic world. (1979 [1947]: 164)

Here is an explicit acknowledgment of what has only recently begun to be
formulated by much of the literature cited above. In phenomenological
terms it suggests the preobjective character of bodily being-in-the-world
and likewise suggests two possible consequences of objectification, that is
the individuation of the psychological self and the instantiation of dualism
in the conceptualization of human being.

In the example from Leenhardt, cultural change in the colonial encounter
reveals the play of the preobjective and objectified body in experience. We
must emphatically not conclude here that the body in "primitive" culture is
necessarily preobjective while the body in "civilized" culture is always
objectified. Objectification is the product of reflective, ideological knowledge, whether it be in the form of colonial Christianity, biological science,
or consumer culture. Our lives are not always lived in objectified bodies, for
our bodies are not originally objects to us. They are instead the ground of
perceptual processes that *end* in objectification (Merleau-Ponty 1962;
Csordas 1990, 1993, 1994), and the play between preobjective and objectified bodies within our own culture is precisely what is at issue in many of the
contemporary critiques.

What most clearly distinguishes the concern with embodiment from the
various forms taken by the anthropology of the body is the methodological
and epistemological problematization of a series of interrelated conceptual
dualities, among which that between the preobjective and objectified is only
the first we have mentioned. Immediately implicated is the conventional
distinction between mind and body, along with a series of derivative distinctions between culture and biology, the mental and the material, culture and
practical reason, gender and sex. It appears at times that there is, among
champions of the body in contemporary human-science theorizing, a tendency to vilify what is usually called "Cartesian dualism" as a kind of moral
abjection. Descartes himself introduced the doctrine as a methodological
distinction, a valuable aid to analysis and a way to free scientific thought
from subjection to theology and strict institutional supervision by the
Church. The philosopher is doubtless not entirely to blame for the ontologi-
zation of the distinction, and the way it has become embedded in our ways of
thinking.[3]

Perhaps the most lucid extended critique of the mind/body duality has been provided by Leder (1990). From a phenomenological standpoint based in the work of Merleau-Ponty and others, he begins with the observation that in everyday life our experience is characterized by the *disappearance* of our body from awareness, describing how the "body not only projects outward in experience but falls back into unexperienceable depths" (ibid.: 53). On the other hand, the vivid but unwanted consciousness of one's body in disease, distress, or dysfunction is a kind of *dys-appearance*, a bodily alienation or absence of a distinct kind: "No longer absence *from* experience, the body may yet surface as an absence, a being-away *within* experience" (1990: 91). Predicated on this analysis, Leder rehabilitates the experiential core of Cartesian dualism, while at the same time identifying its fundamental error. For the dualist, "An experiential disappearance is read in ontological terms. Yet ... this disappearance arises precisely from the *embodied* nature of mind. The body's own structure leads to its self-conceal-ment" (ibid.: 115), and thus to a notion of the immateriality of mind and thought. Meanwhile, alienation from the body as it dys-appears in times of breakdown or problematic operation leads to a "natural bias of attention towards the negative" (ibid.: 127), a bias elaborated in the Western tradition by construing the body as the source of epistemological error, moral error, and mortality. Mind/body dualism is thus identified as a culturally shaped "phenomenological vector," that is "a structure of experience that makes possible and encourages the subject in certain practical or interpretive directions, while never mandating them as invariants" (ibid.: 150).

The example from Leenhardt gives us the body as an important site for analyzing the relationship between the preobjective and the objectified, and Leder's analysis shows how the duality of mind and body calls into question the further distinction between the experiential and the ontologi-cal. Close on the heels of these problematic relations is the perennial problem of the relation between subject and object. The indeterminacy of this relation is highlighted by the observation that, depending on one's methodological standpoint, both mind and body can be construed as either subject or object. Thus mind can be an object, a "central processing mechanism" (Shweder 1990) as it is for cognitive science and mainstream psychology, or it can be the Cartesian subject of rational thought and moral reflection. Body can also be either an object, as it is for contemporary technological medicine and conventional biological science, or it can be the subject of sensation, experience, and world. For anthropology, to under-stand the body as the biological raw material on which culture operates has the effect of excluding the body from original or primordial participation in the domain of culture, making the body in effect a "precultural" substrate. Mind is then invariably the subject and body is an object either "in itself" or one that is "good to think." Little space remains to problematize the alternative formulation of body as the source of subjectivity, and mind as the locus of objectification.

The possibility, arising from the cultural and historical changes outlined at the beginning of this introduction, that the body might be understood as a seat of subjectivity is one source of challenge to theories of culture in which mind/subject/culture are deployed in parallel with and in contrast to body/object/biology. Much of our theorizing is heir to the Cartesian legacy in that it privileges the mind/subject/culture set in the form of representation, whether cast in terms of rules and principles by social anthropology, signs and symbols by semiotic/symbolic anthropology, text and discourse by structural/poststructural anthropology, or knowledge and models by cognitive anthropology. In the human-science literature relevant to cultural theory a critique of representation has begun to take shape. There is both a substantive and an epistemological form taken by this critique. The former is a cultural critique that objects to the ideological substance of representations and seeks more apt ones. The latter is a methodological critique that objects to the dominance of representation as an epistemological modality.

There are several discursive sites for the critique of representation. Feminist theory offers critiques of the way women are represented in terms of body, biology, emotion, sexuality, and instinct (Humm 1990; Suleiman 1986; Jaggar and Bordo 1989; Grosz 1991; Jacobus et al. 1990). Much of the feminist critique comes from disciplines such as literature and philosophy and operates in a poststructuralist semiotic paradigm that questions the content of specific representations while assuming the pragmatic and epistemological primacy of representation. Others challenge the bounds of representation, including existential features of subjectivity within a semiotic paradigm as in Julia Kristeva's (1986) notions of the semiotic chora and *jouissance*, arguing for the existential immediacy of bodily experience (Bigwood 1991), or taking issue with the exclusion of identity and agency in the Foucauldian account of the body (McNay 1991).

A second site of the critique of representation is the philosophy of agency/action. Charles Taylor (1985, 1989), for example, takes issue with a Cartesian theory that identifies subjectivity as internal representation in a "monological" form projected on a "premoral" world, opting instead to construe subjectivity as interpersonal engagement via a "conversational" form within a world constituted by existential concerns. Paul Ricoeur (1991) examines the bounds of representation in his attempt to move from a hermeneutics of text to a hermeneutic of action, and from a semiotic of metaphor to an experiential theory of imagination.

In anthropology the critique of representation has largely taken the form of a critique of ethnographic writing (Clifford and Marcus 1986; Marcus and Fischer 1986; Stoller 1989). The substantive issues in this critique are political and ideological: by what right do we represent the ethnographic other, what are the consequences of doing so, what are the best alternative modes of representation? Occasionally a more radical critique appears of representation as a privileged epistemological modality. From the direction of postmodernism, Tyler (1987: 58) asks "why not reject outright the whole idea of the sensorium, of representation, of the correspondence between inner and outer signifiers whether known as mind and body, thought and language, words and things, or any of the 'othering' dualisms that have trapped us?" He argues that the point of ethnographic "discourse is not to make a better representation, but to avoid representation," suggesting instead that ethnography would do better to "evoke" than to "represent" (ibid.: 205–8). From the direction of phenomenology, Jackson uncovers the representationalist bias in the anthropology of the body itself, particularly in the work of Douglas where "the human body is simply an object of understanding or an instrument of the rational mind, a kind of vehicle for the expression of a reified social rationality" (1989: 123). He argues that the "subjugation of the bodily to the semantic is empirically untenable ... meaning should not be reduced to a sign which, as it were, lies on a separate plane outside the immediate domain of an act" (ibid.: 122). He refers to the methodological standpoint that captures the existential immediacy of bodily existence as "radical empiricism," a term also adopted by Stoller (1989: 151–6) in his phenomenologically oriented effort to develop an evocative anthropology of the senses.

It will not do to identify what we are getting at with a negative term, as something non-representational. We require a term that is complementary as subject is to object, and for that purpose suggest "being-in-the-world," a term from the phenomenological tradition that captures precisely the sense of existential immediacy to which we have already alluded. This is an immediacy in a double sense: not as a synchronic moment of the ethnographic present but as temporally/historically informed sensory presence and engagement; and not unmediated in the sense of a precultural universalism but in the sense of the preobjective reservoir of meaning outlined above. The distinction between representation and being-in-the-world is methodologically critical, for it is the difference between understanding culture in terms of objectified abstraction and existential immediacy. Representation is fundamentally nominal, and hence we can speak of "a representation." Being-in-the-world is fundamentally conditional, and hence we must speak of "existence" and "lived experience."

In general terms, the distinction between representation and being-in-the-world corresponds to that between the disciplines of semiotics and phenomenology. There are without question equally as many variants of one as of the other, and to some extent the representation/being-in-the-world duality reappears within each. Thus within semiotics, broadly conceived there is the tension between text and discourse (Tyler 1987, Lutz and Abu-Lughod 1990), while within phenomenology there is the tension between phenomenology proper and hermeneutics (Ricoeur 1991; Caputo 1986). In anthropology, phenomenology is a poor and underdeveloped cousin of semiotics, and Clifford (1986: 10) does not even mention it among the "proliferating positions" from which interdisciplinary theorizing about the limits of representation has issued.

The dominance of semiotics over phenomenology, and hence concern with the problem of representation over the problem of being-in-the-world, is evident in the relation between the parallel distinction between "language" and "experience." It is still common for those who express interest in the study of experience to confront an objection that runs something as follows: "You cannot really study experience, because all experience is mediated by language – therefore one can only study language or discourse, i.e. representation." I would argue that the polarization of language and experience is itself a function of a predominantly representationalist theory of language. One need conclude neither that language is "about" nothing other than itself, nor that language wholly constitutes experience, nor that language refers to experience that can be known in no other way. One can instead argue that language gives access to a world of experience in so far as experience comes to, or is brought to, language. Ricoeur (1991: 41–2) has pointed to the "derivative character of linguistic meaning . . . It is necessary to say first what comes to language" in processes of presence, memory, and fantasy, in stances such as certitude, doubt and supposition, and in degrees of actuality and potentiality that precede "the properly linguistic plane upon which the functions of denomination, predication, syntactic liaison, and so on come to be articulated." The notion that language is itself a modality of being-in-the-world can be traced at least as far as Herder and Humboldt, and is perhaps best captured in Heidegger's notion that language not only represents or refers, but "discloses" our being-in-the-world.

The dominance of semiotics over phenomenology is also evident in the prominence of the metaphor of textuality in contemporary cultural theory. The essay by Ricoeur (1991 [1971]) on the "model of the text" was pre-eminent in this respect, emphasizing the surpassing of the event by the meaning that constitutes the "paradigmatic function" of texts "with respect to the structuring of the practical field in which individuals figure as agents or as patients" (Ricoeur 1991: xiv, 144–67). Ricoeur did not abandon a concern with being-in-the-world in his influential essay, and in later work reversed his priorities, "allowing the concern with practice to reconquer the preeminence that a limited conception of textuality had begun to obliterate"

(1991: xiv). Anthropologists have by and large not followed this movement, but have tended to elaborate a reading of Geertz's (1973) version of the text metaphor, one that is more explicitly semiotic than Ricoeur's hermeneutic version. Geertz's version of the text metaphor leans toward the representational pole in so far as it is combined with the definition of cultures as systems of symbols and an extrinsic theory of thought that draws out dichotomies between cultural and biological/genetic, and between public and private sources of information. This elaboration has taken place in an intellectual climate influenced by Derrida (1976) and the partisans of deconstruction, who operate under the motto that there is nothing outside the text.

Without going so far as to suggest that the text metaphor has become a representationalist trap for cultural theory (cf. Fernandez 1985), it is in accord with the argument we have developed to place the body in a paradigmatic position complementary to the text rather than allowing it to be itself subsumed under the text metaphor. Already the human science literature is replete with references to the body as a kind of readable text upon which social reality is "inscribed." In such accounts the body is a creature of representation, as in the work of Foucault (1979, 1980), whose primary concern is to establish the discursive conditions of possibility for the body as an object of domination (see also Turner, Chapter 1 in this volume). What about the body as a function of being-in-the-world, as in the work of Merleau-Ponty (1962, 1964), for whom embodiment is the existential condition of possibility for culture and self?

In defining this paradigmatic function it is useful to recall Barthes's distinction between "the work" as a material object that occupies space in a bookstore or on a library shelf, and "the text" as an indeterminate methodological field that exists caught up within a discourse and is experienced as activity and production (1986: 57–58). Instead of Barthes's "work" and "text," I prefer "text" and "textuality," and to them I would like to juxtapose the parallel figures of the "body" as a biological, material entity and "embodiment" as an indeterminate methodological field defined by perceptual experience and mode of presence and engagement in the world. Thus defined, the relation between textuality and embodiment as corresponding methodological fields belonging respectively to semiotics and phenomenology completes our series of conceptual dualities. The point of elaborating a paradigm of embodiment is then not to supplant textuality but to offer it a dialectical partner. That the paradigm of textuality is far ahead of the paradigm of embodiment is without question (see Hanks 1989), but the formulation of their relation promises the grounds for future examination of, for example, the relation between the semiotic notion of intertextuality and the phenomenological notion of intersubjectivity.

Plan of the volume

The expectation that an approach claiming to be grounded in embodiment should be worked out with concrete empirical data is well met in the chapters that follow. Considerable cultural diversity is represented, with authors drawing their arguments from work among Cambodians, Fijians, Chinese, Salvadorans and other Latin American peoples, Euro-Americans, Sri Lankan Tamils, and Navajos. Several of the authors call upon their personal experience as data, not as "introspectionists," but in judicious, and sometimes courageous, use of ethnographic reflexivity. Much of the empirical material comes from experiences of affliction, either in the form of illness, of political violence, or of both. Indeed, a focus on the most vivid exemplars, in this case the modalities of affliction and suffering (see Kleinman and Kleinman 1991), is arguably necessary in the formative stages of an intellectual enterprise. Yet the authors in general are concerned less with affliction *per se* than with contributing to a theory of culture and self grounded in embodiment. It is this concern that has guided my organization of the chapters, an act that is inevitably rhetorical in nature, with consequences for how the volume is perceived. This consideration is all the more relevant when the interests of contributors overlap substantially. For example, the methodological stance of embodiment vis-à-vis biology does not receive its own section, yet is a concern addressed by Lyon and Barbalet, by Jenkins and Valiente, and by Csordas.

Part I consists of a chapter that extends the methodological critique of representation, and another that offers a synthetic argument for integrating embodiment into social theory. Terence Turner renews the work on bodiliness he began over a decade ago (Turner 1980), observing that the body in contemporary capitalist society is a site of both social inequality and personal empowerment. His argument that the appropriation of bodiliness is the fundamental matrix or material infrastructure of the production of personhood and social identity elaborates the notion of the body as existential ground of culture and self, and his distinction between the body as a set of individual psychological or sensuous responses and as a material process of social interaction captures the distinction between body and embodiment outlined above (see also Csordas 1990, 1993). Turner launches a frankly polemical critique of Foucault and poststructuralist theories of the body. He points to the crisis of subjectivity that has led to the prominence in social theory of a passive, representationalist body, and to it juxtaposes a body of being-in-the-world that collapses dualities between subjective and objective, meaningful and material. Beginning at the same historical moment but with a different point in mind than Bourdieu (1988) in his *Homo Academicus*, he identifies poststructuralism as an academic response to the Parisian events of May 1968. As a result, structure was replaced by power, *langue* by *parole*, and mind by body, but all without a corresponding substitution of subject for object. The absence of agency and the possibility for critique in the key concepts of power, discourse, and body leads Turner to define Foucault and his followers not as theorists of the body, but as "anti-bodies." In the wake of 1968 the body is the locus of personal politics, and control of the body is control of the relations of personal production. In conceiving these relations in terms of a body that is inherently plural, existing among other bodies, Turner implicitly offers a link between the political economic notion of relations of production and a phenomenological notion of intersubjectivity as the interactive integument of embodied existence, thus taking a step toward Merleau-Ponty's (1964: 25) unfinished project of linking perceptual reality with cultural and historical analysis.

In Chapter 2, Lyon and Barbalet offer a contrast between two views in contemporary social theory, that of the body as the passive object of ideological representation and as the active subject of embodied being-in-the-world. They note the objectification of the body in consumer culture and in medical practice, and argue that scholarly treatments in large part reflect the ideology embedded in these cultural domains and deny what, in similar vein to Turner, they regard as the intercommunicative and active nature of the body. Going beyond the observation by Scheper-Hughes and Lock (1987) that emotion is the "mediatrix" among the individual body, the social body, and the body politic, Lyon and Barbalet suggest that close attention to the role of emotion in social life can be a corrective to undue objectification, so long as emotion is construed as both embodied and social or relational in its origins and its consequences. Building on an account of emotion in contemporary ethological and evolutionary theory, they emphasize the dual haptic and affective senses of "feeling." They further argue that the interactive and relational character of emotion offers a way for a phenomenologically grounded approach to embodiment to move beyond microanalytic, subjective, internal, individualist analysis toward an open horizon in which social institutions can be understood in terms of their characteristic bodily relations, and embodied agency can be understood as not only individual but institution-making (see also Jenkins and Valiente, Chapter 7 in this volume). The authors point to Merleau-Ponty's phenomenology of perception and Scheler's phenomenology of feeling as fruitful means to this end, means which are empirically elaborated in the contributions to the present volume by Ots and by Csordas.

Part II emphasizes the essentially intersubjective and social nature of bodily experience with respect to themes of form, appearance, and motion. In Chapter 3, Lindsay French examines the political economy of altered body morphology in a camp for Cambodians displaced by their recent civil war, focusing on the presence of substantial numbers of young men who have lost limbs to land mines. From the methodological standpoint which we elaborate in this volume, her work has the merit of examining the body under power and domination without losing the body as subject. Answering a call we made above, and made also in chapters by Turner and by Lyon and Barbalet, she defines her topic not only as the private experience of amputees, but also as public. It is thus a phenomenon constituted as bodily experience not only for the amputees themselves, but for the community which must adjust to a high proportion of its members who are dismembered – not only an intrasubjective experience but an intersubjective transformation of the behavioral environment and its habitus. In placing these concerns against the political economic context of the war, she contributes to the mediation of conceptual dualities by juxtaposing the work of Hallowell on self and Foucault on power. French offers a sophisticated analysis under the concepts of local moral world, power/knowledge, and the political construction of affect, successively examining the amputees' culturally defined sense of losing their capacity, competence and courage. In doing so she is able to balance the relation between the amputees' lived experience of karmic status within a Buddhist habitus, and their position as both marginally productive and abjectly subjected beings within a political ethos.

In Chapter 4, Anne Becker shows both how the social inscribes its values onto the body, and how the body is the ground of the self among Fijians. Whereas in the case described by French, amputation is a phenomenon traumatically forced upon the consciousness of the Cambodian community, among Fijians there is a culturally elaborated somatic mode of attention to body shape, weight gain or loss, and other bodily changes, along with a repertoire of cultural and moral meanings of hunger, appetite, food sharing, and the onset of pregnancy. Becker shows how changes in body morphology index the salient psychocultural theme of *care*, intriguingly reminiscent of Heidegger's notion of existential care (*Sorge*). Her description of how Fijians closely monitor changes in body shape combines aspects of textuality and embodiment, including both a sense of the reading of bodies as texts, and an intersubjective somatic mode of attending to others, grounded in the sensory determination of care. Finally, Becker compares the Fijian idealization of body shape with that common in Western cultures. The West, as shown in feminist critique and the critique of consumer culture, cultivates the body as a representation of self, hence alienating body and self. Fijians, however, cultivate one another's bodies as a group rather than a personal endeavor, such that the locus of collective representation is the changes wrought through the care of others.

Building in part on the work of Max Scheler, Thomas Ots points out in Chapter 5 that the very term "embodiment" can be misleading if it is understood as referring not to an existential condition but to a process of putting culture or mind into a body that is objectified and thinglike. Instead he opts for the German term *Leib*, the live-body-self-subject for which no equivalent exists in English. Ots casts the relation between representation and being-in-the-world in Scheler's terms of the relation between mind and life, and their reconciliation in the "enlivenment of the mind." He uses these ideas to frame an analysis of *qigong*, a cathartic healing movement in the contemporary Peoples' Republic of China. In *qigong* practice bodily spontaneity is thematized and objectified in a cultural context where spontaneous movements are problematized in the face of cultural values on quietness, relaxation, and harmony in conjunction with a repressive political atmosphere. In the letters and poems of practitioners Ots encounters an exceedingly rich and *leibly* cultural phenomenology of movement, sensation, metaphor, and emotional transmutation.

The chapters in Part III share a remarkable success in suspending cultural accounts of bodily experience in the indeterminate space between the analytics of representation and being-in-the-world. In particular, Chapters 6 and 7 lend additional substance to Kirmayer's (1992) insight that metaphor is the critical meeting ground between textuality and embodiment. In her contribution, Setha Low problematizes the relation between mind and body, sensation and sense, and biology and culture in the embodied metaphor of *nervios* or "nerves" across five different cultural settings. Surveying data from Costa Rica, Puerto Rico, Guatemala, Newfoundland, and Eastern Kentucky, she argues that the lived experience of *nervios* and the cross-cultural variety of "senses of the body" among *nervios* sufferers correspond to sociopolitical and cultural conditions of distress. Low's concern is for how *nervios* varies as an embodiment of distress across the various cultures she examines, and for how the body is thus a mutable mediator between self and society. However, her notion of *nervios* as an "embodied metaphor" refers not to a metaphor about the body or one that is imposed upon the body, but to a metaphor that is emergent in bodily experience. In elaborating this notion she outlines the analytic meeting point between understandings of the body as a source of meaning and as a representation of social forces.

In Chapter 7 Janis Jenkins and Martha Valiente take this analysis to a greater level of specificity by examining political, bodily, emotional, and psychopathological dimensions of one of the sensations typically associated with *nervios*/nerves among Salvadoran women refugees to the United States, that of intense heat or *calor*. Low had already observed in her contribution that if *nervios* is an embodied metaphor of distress, each of the sensations associated with it can also be understood as metaphors, either of nerves as a global condition or, more directly, of distress. In this chapter

Jenkins and Valiente show how *calor* itself is a complex notion that can be described with vivid metaphoric language by the afflicted, and in doing so they further problematize the sensation/representation contrast. They identify at least three features of indeterminacy. First, *calor* is unevenly objectified in explicit cultural terms, some people recognizing it by name and others describing the experience without categorizing it as a general type of phenomenon. Second, from the women's narratives the authors distill an analysis of polytropy, the use of multiple figures of speech, as evidence of the essential indeterminacy of the experience across narrative accounts and categories of psychiatric diagnosis. Third, by placing it in the sociopolitical context of domestic violence and the violence of *la situacion* in the Salvadoran civil war, they show how *calor* is existentially isomorphic with, and not a representation of, anger and fear. Jenkins and Valiente argue for consideration of the body as a generative source of culture rather than as a *tabula rasa* upon which cultural meaning is inscribed. They conclude by identifying a series of methodological assumptions that are thrown into question from the standpoint of embodiment.

While adding significantly to the theme of inseparability of bodily experience and cultural meaning elaborated by Low and by Jenkins and Valiente, Carol Laderman's approach to food and self in Chapter 8 should also be read in light of Becker's discussion in Chapter 4, which was oriented around the notion that cultural values are encoded in body morphology. For Laderman it is the practices of food ingestion and avoidance that are of concern as she describes the relation between sensory reality and symbolic structure in East Malay culture. First, Laderman analyzes Malay hot and cold humoral reasoning with regard to illness attributed to superheated spirit attacks or humoral imbalance, and humoral effects of diet in the causation and treatment of illness. Unlike other treatments of humoral reasoning, Laderman's includes a sensory component, including self-perceptions of metabolic changes induced by ingesting certain kinds of foods. Second, she examines the concept of *bisa*, which combines the meanings of "poison" and "power," showing how it articulates food-avoidance practices pertaining both to physiological and symbolic danger. Third, she explicates the Malay notions of *semangat* (Spirit of Life) and *angin* (Inner Winds) as keys to understanding the intellectual, cosmological, emotional, sensible, and temperamental dimensions of the Malay self. Fourth, she offers an intriguing reflective description of her own incorporation of elements of Malay embodiment in her reactions to humorally hot or cold foods, the experience of Inner Winds during trance, and the flight and return of her Spirit of Life during an illness. She concludes with a reflection on the relation between representation and being-in-the-world in terms of the mooring of symbolic systems in the experiential world.

The contributors to Part IV each deal in their own way with the relation between language and bodily experience by struggling to articulate the essentially mute preobjective world of pain. In Chapter 9, Jean Jackson takes up the problematic of subject/object, mind/body, and language/ experience in the context of one of the greatest challenges for a theory of culture and self grounded in embodiment, namely the medical condition of chronic pain. Based on a study of patients in a specialized inpatient pain treatment center in New England, she carefully sorts out the inherently indeterminate relations among pain behavior, the experience of pain, and the emotional states accompanying pain. She examines the dialectic of subjectification and objectification as an existential struggle over the "reality" of pain and its experience as self or not-self. She vividly illustrates the cultural immediacy of the Cartesian duality (Leder 1990) in patients' attempts to define control in terms of mind over matter or matter over mind. Jackson pays close attention to language and the communication of pain-experience between patients and non-sufferers and among patients themselves. Distinguishing carefully between pain and its causes, physical and emotional pain, and pain as sensation and emotion, she highlights the existential shock for her afflicted interlocutors of moving back and forth between the "pain-full world" and the world of everyday life.

In Chapter 10, E. Valentine Daniel turns our attention from the author-less pain of the medical patient to the intentionally inflicted pain of political torture. The juxtaposition of these chapters highlights the tragic irony that the social origin of the latter form of pain does not render it less, but perhaps more unrepresentable. Daniel examines the experience of people tortured during the ongoing vicious civil war in Sri Lanka, and through their pain traces the limits of representation in semiotic anthropology. Unlike Jackson's sufferers from chronic pain who were able to develop a community of meaning, Daniel sees in the affectively flat memory of pain among the tortured, and their unwillingness to accept that others have been tortured, their experience of the "sheer worthlessness of all attempts to communicate something that was so radically individuated and rendered unshareable." He identifies beauty, as the preeminent example of what Peirce called "qualisigns," as a semiotic first with the felt quality of prereflective, immediate, uncategorized experience. Pain, on the other hand, is a "sinsign" in which the immediacy of semiotic firstness is overwhelmed by the alien-ated otherness of secondness. He goes on to identify terror and art as cultural forms for objectifying pain, allowing the opening out into semiosis of experience that had been "compacted in pain." In a vivid portrayal of his tortured interlocutors' ascent from speechlessness to the ability to form elementary metaphors, he captures the experiential trajectory of objectifi-cation that outlines the existential moment of relation between representation and being-in-the-world, and the theoretical moment of relation between semiotics and phenomenology.

In Chapter 11, Cathy Winkler analyzes another form of pain as a result of political violence, rape. Drawing courageously on her personal experience of this trauma, she examines the mind/body dissociation consequent on rapists' "forcefully inserting land mines of emotional upheaval into the bodies of their victims." Her argument underscores the urgent need for a paradigm of embodiment in the era of Anita Hill and Clarence Thomas, Desiree Washington and Mike Tyson. Here is the struggle between one's own self-objectification and the objectification of self by the violence of others. Unlike Daniel, who could only observe his Tamil interlocutors' struggle for semiosis, Winkler as "investigator-victim" is able to deal with her ethnographic informants as "victim-investigators." In a unique, inverted ethnographic relationship, the anthropologist provides her informants with an analytic framework, and they take responsibility for editing the texts of their interviews. Beginning with a critique of the symptoms and stages abstracted as "rape trauma syndrome," Winkler identifies a series of five overlapping contexts of trauma that define the cultural and existential meaning of rape. The existential ground of culture, in particular of emotional meaning, becomes especially vivid in the body's visceral, prereflective recognition of the rapist, even while the victim is unable to evoke a definite visual/representational memory of his face.

Finally, Chapter 12 is my attempt at a cultural phenomenology of the physical and existential pain suffered by a young Navajo man afflicted with a tumor-induced seizure disorder. Framed by reflections on Heidegger and Merleau-Ponty, it is also an attempt to anticipate two potential objections to a phenomenological paradigm of embodiment: first, that the domain of preobjective experience is somehow "precultural;" and second, that "embodiment" throws out the biology with the bathwater. The young man's strategy for making sense of his affliction combines elements of Navajo Peyotist healing and spirituality with elements of Anglo-American explanatory models and biomedical treatment, and presupposes a complex interaction between biology and culture. I analyze the patient's narrative in order to show the cultural formation of preobjective experience, and that language can be understood not only as representation but as disclosure of reality and being-in-the-world. I also show the bodily immediacy in which schemas of contamination by lightning, the ritual number four, and being shot through with an object by witchcraft come into play. Phenomenology and neurology are brought into dialogue with a discussion of how the behavioral syndromes associated with temporal-lobe lesions and the effects or ritual peyote ingestion derive existential meaning in the context of a bodily synthesis of language, thought, religious experience, and healing.

Given the early stage of development of a paradigm of embodiment, it should come as no surprise that the contributions to this volume do not adhere strictly to any one paradigmatic position. The reader will note variations in use of the term embodiment itself: most authors regard it as an existential condition, others as a process in which meaning is taken into or upon the body, yet others prefer the term bodiliness over embodiment. Nevertheless, problematizing the body and embodiment places each author within the nexus of dualities I have elaborated above, to work out his or her own position with respect to the relations between preobjective and objectified, mind and body, subject and object, representation and being-in-the-world, semiotics and phenomenology, language and experience, textuality and embodiment. Their collective assertion is that these pairs of terms define a critical moment in theorizing about culture and self, and further that although none of these dualities is spurious, neither are the polar terms irreconcilably opposed. We are well reminded, for example, of Peirce's inclusion of habit in his semiotics and Merleau-Ponty's concern with signs in his phenomenology. In this light our purpose is to identify the terrain on which opposed terms meet, whether they are understood to remain in tension or to collapse upon one another. That terrain is marked by the characteristic reflectiveness and the process of objectification that define human consciousness, giving substance to representation and specificity to being-in-the-world.

NOTES

1 For different perspectives and thorough bibliographic reviews see Frank (1991, B. Turner (1991), and Lock (1994).
2 Lakoff (1987) and Johnson (1987) have argued that many of the metaphors that structure our experience are derived from body-based image schemas. They suggest that the body and its inherent orientations are "taken up" into culture, becoming "the body in the mind," without attempting to account for the reciprocal sense in which one can simultaneously speak of "the mind in the body." Despite its intent their approach thus entertains a complex flirtation with reductionism, dualism, and intellectualism. Following Morleau-Ponty, I would argue that the body is always already cultural, and that rather than asking how metaphors instantiate image schemas it is more apt to begin with the lived experience from which we derive image schemas as abstract products of analytic reflection. By the same token, Quinn's (1991) critique of Lakoff and Johnson to the effect that culture takes priority over the body does no more than invert their argument, and is thus misplaced by presuming a distinction between body and culture that a priori excludes the bodily from the cultural.
3 In anthropology, the mind/body separation is cast predominantly in terms of a dichotomy between culture and biology. Not only is this dichotomy institutionalized in the distinction between cultural and physical anthropology, but within cultural anthropology reference to the body has, until recently, tended to be synonymous with an invocation of biology (Csordas 1990).

REFERENCES

Barthes, Roland, (1986) *The Rustle of Language*, R. Howard, trans. New York: Hill and Wang.

Bigwood, Carol (1991) Renaturalizing the Body (With a Little Help from Merleau-Ponty). In Elizabeth Grosz, ed., *Feminism and the Body*, Special issue of *Hypatia* 6 (3): 54–73.

Bordo, Susan (1990) Reading the Slender Body. In Mary Jacobus, Evelyn Fox Keller, and Sally Shuttleworth, eds., *Body/Politics: Women and the Discourses of Science*. New York: Routledge, pp. 83–112.

Bourdieu, Pierre (1988) *Homo Academicus*. Stanford: Stanford University Press.

Bynum, Caroline Walker (1989) The Female Body and Religious Practice in the Later Middle Ages. In Michel Feher, ed., *Fragments for a History of the Human Body, Part I*. New York: Zone, pp. 160–219.

Caputo, John D. (1986) Husserl, Heidegger, and the Question of "Hermeneutic" Phenomenology. In Joseph Kockelmans, ed., *The Companion to Martin Heidegger's "Being and Time."* Washington: Center for Advanced Research in Phenomenology and University Press of America, pp. 104–26.

Clifford, James (1986) Introduction. In James Clifford and George E. Marcus, eds., *Writing Culture: The Poetics and Politics of Ethnography*. Berkeley: University of California Press, pp. 1–26.

Clifford, James and George E. Marcus eds. (1986) *Writing Culture: The Poetics and Politics of Ethnography*. Berkeley: University of California Press.

Csordas, Thomas J. (1990) Embodiment as a Paradigm for Anthropology. *Ethos* 18: 5–47.

(1993) Somatic Modes of Attention. *Cultural Anthropology* 8: 135–56.

(1994) *The Sacred Self: A Cultural Phenomenology of Charismatic Healing*. Berkeley: University of California Press.

Derrida, Jacques (1976) *Of Grammatology*. Baltimore: Johns Hopkins University Press.

Desjarlais, Robert (1992) *Body and Emotion: The Aesthetics of Illness and Healing in the Nepal Himalayas*. Philadelphia: University of Pennsylvania Press.

Devisch, Rene and Antoine Gailly, eds. (1985) *Symbol and Symptom in Bodily Space-Time*. Special issue of the *International Journal of Psychology* 20: 389–663.

Doane, May Ann (1990) Technophilia: Technology, representation, and the Feminine. In Mary Jacobus, Evelyn Fox Keller, and Sally Shuttleworth, eds., *Body/Politics: Women and the Discourses of Science*. New York: Routledge, pp. 163–76.

Douglas, Mary (1973) *Natural Symbols*. New York: Vintage.

Featherstone, Mike (1991) The Body in Consumer Culture. In Mike Featherstone, Mike Hepworth, and Bryan S. Turner, eds., *The Body: Social Process and Cultural Theory*. London: Sage Publications, pp. 170–96.

Featherstone, Michael and Mike Hepworth (1991) The Mask of Ageing and the Postmodern Life Course. In Mike Featherstone, Mike Hepworth, and Bryan S. Turner, eds. *The Body: Social Process and Cultural Theory*. London: Sage Publications, pp. 371–89.

Feher, Michel (1989) Introduction. In Michel Feher, ed., *Fragments for a History of the Human Body, Part I*. New York: Zone, pp. 10–17.

Feldman, Allen (1991) *Formations of Violence: The Narrative of the Body and Political Terror in Northern Ireland*. Chicago: University of Chicago Press.

Fernandez, James (1985) Exploded Worlds: Text as a Metaphor for Ethnography. *Dialectical Anthropology* 10: 15–26.

(1990) The Body in Bwiti: Variations on a Theme by Richard Werbner. *Journal of Religion in Africa* 20: 92–111.

Foucault, Michel (1979) *Discipline and Punish: The Birth of the Prison.* New York: Vintage.

(1980) *The History of Sexuality. Volume I: An Introduction.* New York: Vintage.

(1986) *The Care of the Self. The History of Sexuality. Volume III.* New York: Vintage.

Frank, Arthur (1991) For a Sociology of the Body: An Analytical Review. In Mike Featherstone, Mike Hepworth, and Bryan S. Turner, eds., *The Body: Social Process and Cultural Theory.* London: Sage Publications, pp. 36–102.

Frank, Gelya (1986) On Embodiment: A Case Study of Congenital Limb Deficiency in American Culture. *Culture, Medicine, and Psychiatry* 10: 189–219.

Geertz, Clifford (1973) *The Interpretation of Cultures.* New York: Basic Books.

Good, Byron (1954) *Medicine, Rationality, and Experience: An Anthropological Perspective.* Cambridge: Cambridge University Press.

Good, Mary-Jo, Paul Brodwin, Byron Good, and Arthur Kleinman (1992) *Pain as Human Experience: An Anthropological Perspective.* Berkeley: University of California Press.

Gordon, Deborah (1990) Embodying Illness, Embodying Cancer. *Culture, Medicine and Psychiatry* 14: 275–97.

Grosz, Elizabeth, ed. (1991) *Feminism and the Body.* Special issue of *Hypatia* 6 (3).

Hanks, William (1989) Text and Textuality. *Annual Review of Anthropology* 18: 95–127.

Haraway, Donna (1990) Investment Strategies for the Evolving Portfolio of Primate Females. In Mary Jacobus, Evelyn Fox Keller, and Sally Shuttleworth, eds., *Body/Politics: Women and the Discourses of Science.* New York: Routledge, pp. 139–62.

(1991) *Simians, Cyborgs, and Women: The Reinvention of Nature.* New York: Routledge.

Humm, Maggie (1990) *The Dictionary of Feminist Theory.* Columbus: Ohio State University Press.

Jackson, Michael (1989) *Paths toward a Clearing: Radical Empiricism and Ethnographic Inquiry.* Bloomington: Indiana University Press.

Jacobus, Mary (1990) In Parenthesis: Immaculate Conceptions and Feminine Desire. In Mary Jacobus, Evelyn Fox Keller, and Sally Shuttleworth, eds., *Body/Politics: Women and the Discourses of Science.* New York: Routledge, pp. 11–28.

Jacobus, Mary, Evelyn Fox Keller, and Sally Shuttleworth, eds. (1990) *Body/Politics: Women and the Discourses of Science.* New York: Routledge.

Jaggar, Alison M. and Susan R. Bordo, eds. (1989) *Gender/Body/Knowledge: Feminist Reconstructions of Being and Knowing.* New Brunswick: Rutgers University Press.

Johnson, Mark (1987) *The Body in the Mind: The Bodily Basis of Meaning, Imagination, and Reason.* Chicago: University of Chicago Press.

Kant, Immanuel (1978) [1800] *Anthropology From a Pragmatic Point of View*, Victor Lyle Dodell, trans. Carbondale IL: Southern Illinois University Press.

Keller, Evelyn Fox (1990) From Secrets of Life to Secrets of Death. In Mary Jacobus, Evelyn Fox Keller, and Sally Shuttleworth, eds., *Body/Politics: Women and the Discourses of Science*. New York: Routledge, pp. 177–91.

Kirmayer, Laurence J. (1989) Mind and Body as Metaphors. In M. Lock and D. Gordon, eds., *Biomedicine Examined*. Dordrecht: Kluwer Academic Publishers, pp. 57–94.

(1992) The Body's Insistence on Meaning: Metaphor as Presentation and Representation in Illness Experience. *Medical Anthropology Quarterly* 6: 323–46.

Kleinman, Arthur (1988) *The Illness Narratives: Sufferings, Healing, and the Human Condition*. New York: Basic Books.

Kleinman, Arthur and Joan Kleinman (1991) Suffering and its Professional Transformation: Toward an Ethnography of Experience. *Culture, Medicine, and Psychiatry* 15: 275–302.

Kristeva, Julia (1986) *The Kristeva Reader*, Toril Moi, ed. New York: Columbia University Press.

Kroker, Arthur and Marilouise Kroker (1987) *Body Invaders: Panic Sex in America*. New York: St. Martin's Press.

Lakoff, George (1987) *Women, Fire, and Dangerous Things: What Categories Reveal about the Mind*. Chicago: University of Chicago Press.

Leder, Drew (1990) *The Absent Body*. Chicago: University of Chicago.

Leenhardt, Maurice (1979) [1947] *Do Kamo: Person and Myth in a Melanesian World*, Basia Miller Gulati, trans. Chicago: University of Chicago Press.

Lock, Margaret (1993) The Anthropology of the Body. *Annual Review of Anthropology*, 22:133–53.

Lock, Margaret and Pamela Dunk (1987) My Nerves are Broken. In D. Coburn, C. D'Arcy, G. Torrance, and P. New, eds., *Health in Canadian Society: Sociological Perspectives*. Toronto: Fithenry and Wimbleside, pp. 2995–3013.

Lutz, Catherine A. and Lila Abu-Lughod, eds. (1990) *Language and the Politics of Emotion*. New York: Cambridge University Press.

McNay, Lois (1991) The Foucauldian Body and the Exclusion of Experience. In Elizabeth Grosz, ed., *Feminism and the Body*. Special issue of *Hypatia* 6 (3): 125–39.

Marcus, George and Michael M. J. Fischer (1986) *Anthropology as Cultural Critique: An Experimental Moment in the Human Sciences*. Chicago: University of Chicago Press.

Martin, Emily (1992) The End of the Body? *American Ethnologist* 19: 121–40.

Mauss, Marcel (1950) Les Techniques du Corps. *Sociologie et Anthropologie*. Paris: Presses Universitaires de France.

Merleau-Ponty, Maurice (1962) *Phenomenology of Perception*, James Edie, trans. Evanston, IL: Northwestern University Press.

(1964) *The Primacy of Perception*, James M. Edie, ed. Evanston, IL: Northwestern University Press.

Murphy, Robert F. (1987) *The Body Silent*. New York: Henry Holt.

O'Neill, John (1985) *Five Bodies: The Shape of Modern Society*. Ithaca: Cornell University Press.

Ots, Thomas (1991) Phenomenology of the Body: The Subject–Object Problem in Psychosomatic Medicine and Role of Traditional Medical Systems. In Beatrix Pflederer and Gilles Bibeau, eds., *Anthropologies of Medicine: A Colloquium of*

West European and North American Perspectives. Special issue of *Curare.* Wiesbade: Wieweg, pp. 43–58.

Pandolfi, Mariella (1991) Memory Within the Body: Women's Narrative and Identity in a Southern Italian Village. In Beatrix Pflederer and Gilles Bibeau, eds., *Anthropologies of Medicine: A Colloquium of West European and North American Perspectives.* Special issue of *Curare.* Wiesbade: Wieweg.

Quinn, Naomi (1991) The Cultural Basis of Metaphor. In James Fernandez, ed., *Beyond Metaphor: The Theory of Tropes in Anthropology.* Stanford: Stanford University Press, pp. 56–93.

Ricoeur, Paul (1991) *From Text to Action: Essays in Hermeneutics II,* Kathleen Blamey and John B. Thompson, trans. Evanston, IL: Northwestern University Press.

Roseman, Marina (1991) *Healing Sounds from the Malaysian Rainforest: Temiar Music and Medicine.* Berkeley: University of California Press.

Scarry, Elaine (1985) *The Body in Pain: The Making and Un-Making of the World.* New York: Oxford University Press.

Scheper-Hughes, Nancy (1992) *Death without Weeping: The Violence of Everyday Life in Brazil.* Berkeley: University of California Press.

Scheper-Hughes, Nancy and Margaret Lock (1987) The Mindful Body: A Prolegomenon to Future Work in Medical Anthropology. *Medical Anthropology Quaterly* 1: 6–41.

Shweder, Richard (1990) Cultural Psychology: What Is It? In J. Stigler, R. Shweder, and G. Herdt, eds., *Cultural Psychology: Essays on Comparative Human Development.* Cambridge: Cambridge University Press, pp. 1–47.

Stoller, Paul (1989) *The Taste of Ethnographic Things: The Senses in Anthropology.* Philadelphia: University of Pennsylvania Press.

Suleiman, Susan Rubin, ed. (1986) *The Female Body in Western Culture.* Cambridge, MA: Harvard University Press.

Taylor, Charles (1985) The Person. In M. Carrithers, S. Collins, and S. Lukes, eds., *The Category of Person: Anthropology, Philosophy, History.* New York: Cambridge University Press, pp. 257–81.

  (1989) *Sources of the Self: The Making of the Modern Identity.* Cambridge: Harvard University Press.

Turner, Bryan (1991) Recent Developments in the Theory of the Body. In Mike Featherstone, Mike Hepworth, and Bryan S. Turner, eds., *The Body: Social Process and Cultural Theory.* London: Sage Publications, pp. 1–35.

Turner, Terence (1980) The Social Skin. In J. Cherfas and R. Lewin, eds., *Not Work Alone.* London: Temple Smith.

Tyler, Stephen A. (1987) *The Unspeakable: Discourse, Dialogue and Rhetoric in the Postmodern World.* Madison: University of Wisconsin Press.

Vernant, Jean-Pierre (1989) Dim Body, Dazzling Body. In Michel Feher, ed., *Fragments for a History of the Human Body, Part I.* New York: Zone, pp. 18–47.

Part I

Paradigms and polemics

1 Bodies and anti-bodies: flesh and fetish in contemporary social theory

Terence Turner

The current salience of bodily-related political movements and the importance of the body in current social and cultural theorizing is associated with the prominence of body-related themes in the culture of contemporary late capitalism. The body, or embodied subject, is the object of seduction by advertising, interpellation by semiotically loaded commodities, torture by a broad spectrum of political regimes, bitter conflicts over reproductive rights and health care, struggles for the revaluation of alternate sexual identities, threats from new epidemic diseases, and the object of new technologies permitting the alteration of physical attributes hitherto accepted as naturally determined, including cosmetic surgery, asexual and extra-bodily fertilization, multiple forms of intervention in the biological process of reproduction, the modification of genetic traits, and the artificial prolongation or curtailment of the span of life itself. In these and other ways, the body seems thrust into ever-increasing prominence as the arena of social conflicts and repressive controls, as well as some of the most liberating aspects of contemporary culture and social life.

The hypertrophy of commodity production in late capitalism has created an unprecedented emphasis on the plasticity of personal identity, and with it, willy nilly, an emphasis on the ability of people to produce themselves as social identities, taking over the far more sociocentric and narrowly regimented forms of the production of personhood which up until a few decades ago still inhered in traditional family, community, educational and class structures. By one of the neater ironies of the history of capitalism, however, the triumph of individualistic consumerism, and its crowning social achievement, the creation of a socially guaranteed personal space in which individual consumers can produce their own identities, has unleashed a new collective politics of identity, and an equally collective projection of individual bodily concerns in the form of environmentalist and "green" movements implicitly based in large part on Marx's premise that "Nature is man's inorganic body ... [that] man lives on Nature means that Nature is his body, with which he must remain in continuous interchange if he is not to die" (Marx 1964: 112). As the expansion of social and cultural freedom to create one's own identity has pushed against the rigidities of traditional patriarchal, sexual, political and economic forms of domination and control, a new politics of personal empowerment and emancipation has been born to challenge these persisting limitations on personal freedom, starting with the basic power to appropriate one's own body, including its sexuality and reproductive powers, to produce the identity, the social persona or in the newly pervasive phrase, "life-style" one chooses. The inevitable consequence of this has been that bodily qualities such as youth, slimness, fitness, etc. have acquired new importance as tokens of class and status distinctions, as Bourdieu, for example, has prominently argued (Bourdieu 1984).

This new "life politics" (Giddens 1991) of personal identity has focused so largely on issues of bodiliness because, I suggest, the appropriation of bodiliness, in all its aspects, from sexuality and reproductive capacities to sensory powers and physical health, strength, and appearance, is the fundamental matrix, the material infrastructure, so to speak, of the production of personhood and social identity. What is at stake in the struggle for control of the body, in short, is control of the social relations of personal production. Consistently with this, the body remains the site of some (although by no means all) of the most fundamental forms of social inequality and control in contemporary society, as well as some of its characteristic forms of mystified social consciousness. The focus of both identity production and some repressive social controls on the body tends to obscure their social dimensions and exacerbate the tendency to see them and the body itself in individualistic (and often psychologistic) terms.

The emergence of the body as a focus of these and other contemporary historical developments largely accounts for its new prominence in current intellectual culture, social, and psychological theory. More importantly, and more relevantly for the purposes of the present discussion, it also helps to account for some of the major limitations and distortions of the nature of the body as a biological, social, and cultural phenomenon in current social and cultural theory. These distortions may be summed up as, first, a pervasive tendency to ignore or misrecognize the social nature of the body, and the multifold ways it is constituted by relations with other bodies, in favor of a reified conception of the body as bounded individual; and second, a propensity to ignore the primary character of the body as material activity in favor of an emphasis on the body as a conceptual object of discourse. The severance of the body's social roots, its de-materialization as a figment of discourse, and its reification as a transcendental individual combine to promote a general tendency towards the psychologization of discourses on the body. In all of these respects, there is a substitution of "the body" conceived as a set of individual psychological or sensuous responses and needs for the body as material process of social interaction. This substitution betrays much of the contemporary discourse on the body as an expression of the individualistic social ideology of the middle-class professional intellectuals who have developed it as an alternative to class- and other socially based political and intellectual perspectives.

In these key respects the contemporary discourse on the body has emerged as one of the major manifestations of a crisis in the intellectual politics and epistemology of Western social thought, or more precisely, of Western social thinkers and the professional–managerial class to which they belong (Ehrenreich and Ehrenreich 1979; Pfeil 1990: 98–101). This crisis proceeds from a sense of the irrelevance, futility, or discreditedness of fundamental Western political and social institutions, and their associated moral and cultural values, at least from the perspective of the social position of the thinkers in question. Proximate causes of this crisis include the erosion of historic forms of political and social organization (resulting from both the progressive centralization of political and economic power at trans-national levels and the collapse of the eurocentric imperialist world order) and the inadequacy of traditional notions of culture, unified subjective consciousness, and aesthetic canons to deal with the consumerist culture of late capitalism. Many of the assumptions about the interdependence of the individual and society have consequently been called into question, as have the associated assumptions about the meaning of cultural forms, the relation of subjective consciousness to social reality, and the possibility or value of political action, on which civil society and the state, as well as the major traditions of social thought and philosophy in the West, have been based. Significantly, however, this calling into question has tended not to extend to the essential individualism and commitment to notions of knowledge, culture, and consciousness as essentially unconditioned by material social relations or class position which are characteristic of the professional-academic-class perspective of the questioning intellectuals themselves.

The themes, forms and limitations of this theoretical and ideological reorientation of intellectuals to contemporary capitalist society are evident in the current spate of discourses on "the body." One of the characteristic symptoms of this crisis is the development of new forms of social and cultural thought based on the rejection of subjectivity and any epistemological claim of access, however mediated, to objective social or historical reality. These obviously interdependent moves are typically linked to a rejection of systematic social theory, and indeed of any coherent notion of society at all, and along with it the ideal of a theoretically grounded critique of social consciousness and institutions that has been the guiding aspiration of traditional Western social thought. One of the characteristic gambits within this new drift of ostensibly post-political, post-social consciousness has been the attempt to banish "the subject" as a metaphysical entity endowed with relatively autonomous capacities for action and critical consciousness, by replacing it with "the body," conceived for social purposes as a material, but more or less passive, object of disciplines and representations, nevertheless imbued with private, infra-social, crypto-subjective desires and proclivities, usually of a sexual or more broadly psychodynamic nature.

The elevation of the body to the place occupied by subject, agent and social individual in older forms of Western social thought, notwithstanding its apparently "materialist" character as a substitution of a concrete physical entity for an abstract metaphysical concept, has generally involved in practice a focus on conceptual or linguistic representations of the body and an indifference to the body as an objective physical reality. In other words, in much recent theorizing "the body" tends to undergo a subtle transformation from physical object or material activity to a series of discourses conceived in a neo-structuralist manner as autonomous semiotic phenomena. This crypto-structuralist move (often enough accompanied by fierce disavowals of structuralist affiliation) is typically accompanied by a reductionist demotion of social representations or discourses of the body to the status of inauthentic or deceptive manipulations, reserving for individual psychobiological motives and desires whatever quality of authenticity or politico-moral valuation survives the exercise. The political effect is to elevate individual, psychological (above all sexual) motives, at the expense of social, political-economic, and ideological issues and values as the focal considerations in the construction of social identity and what has come to be loosely referred to as "resistance."

In these respects, much of the new discourse on the body and bodiliness, in spite of its apparently similar concerns, has tended to subvert or contradict the approaches of overtly political movements and forms of cultural critique that have contested bodily-related forms of social and cultural oppression based on gender, sexual orientation, race, or ethnicity. Such movements (various feminist tendencies, some gay and lesbian and some black writers and movements, for example) have sought to confront the material social and ideological appropriation of aspects of bodiliness with programs of political action and cultural criticism based on the assertion and advocacy of subjective identity and critical consciousness.

A central rhetorical move of the more influential tendencies within the new body-oriented theorizing is to effectively abstract issues of bodily representation, sexuality, the definition and care of illness, and so on from the specific political-economic and sociocultural dynamics of capitalism (or other historically relevant sociological context) by ascribing them to vague and unspecified trans-historical forces (e.g., "power," "discipline," etc.).

This has tended to subvert the possibility of effective political resistance or socially grounded cultural critique of such body-related social practices. The critical insight that the proliferating commodification of all aspects of bodiliness in contemporary society is driven by the systemic requirements of the accumulation of capital under the specific market conditions of late capitalism, for instance, finds no place in poststructuralist, Lacanian or deconstructionist discourses on the body and its oddly persistent *doppelganger*, the non-existent subject. The issues raised in the confrontation between the contemporary theoretical discourses on the body derived from these and allied tendencies and a genuinely political politics of gender, sexual, racial, or ethnic identity obviously turn on the nature of the social realities that underlie and determine the cultural and phenomenological realities of body and representation alike, just as they condition the possibilities for subjective consciousness and political action. The body has thus become one of the main battlegrounds on which the struggle to forge a critical perspective adequate to the changing features of contemporary social, political, and cultural reality is being fought.

Bodies and anti-bodies in contemporary social theory

The crisis in contemporary Western social and cultural thought that has come, for various reasons, to focus on issues involving the body and bodiliness is manifested in many strands of European and American social thought and cultural theory. In spite of great differences of style, historical sources, and technical vocabulary, these can be seen to converge on certain fundamental theoretical points. Of these, some fall within and some outside the purview of what is normally included under the rubric of "postmodern" thought. Although I shall be concerned in this essay mainly with one tendency within postmodernism, namely poststructuralism, I want to emphasize that I am not concerned here specifically or exclusively with the canonical position of postmodernism. It is important to recognize *both* that the poststructuralist and deconstructionist tendencies in French and American cultural criticism, aesthetic and literary theory, and post-Nietzschean and crypto-pragmatic philosophy overtly identified as postmodernist are not the only game in town, *and also* that a number of the other games not normally assimilated under the same rubric are being played by what look very like some of the same basic rules.

I would now like to try to come to critical grips with some of these rules, as they bear specifically on the interpretation of the social and cultural meaning of the body, through a critical examination of some aspects of the work of Michel Foucault, whose later work contains arguably the most prominent and influential instance of the turn to the body as a primary site of social and cultural theorizing.

Foucault: from structuralist episteme to poststructuralist body

The reception of French social theories in the English-speaking world routinely involves their reinterpretation as relatively academic exercises in the English or American sense. They are dissociated from their original context in a political culture where moves are (or until quite recently still were) calibrated in relation to the cardinal points of Marxist left and crypto-Fascist right and overshadowed, within the last half-century, by the great collective traumas of national military defeat and foreign political domination, first by hostile (German) occupiers and then by victorious (Anglo-Saxon) allies. They are also dissociated from their roots in an intellectual culture whose formal hierarchization and centralization in Paris has promoted a form of personalized intellectual competition without parallel in Anglo-Saxon experience. This cross-Channel or trans-Atlantic transformation can result in considerable if unwitting changes in the meanings and functions of key concepts and terms.

Structuralism is a case in point. Originating in Lacan's seminars in the late 1930s and Lévi-Strauss's early work in the late 1940s, it developed into an important movement in the 1950s and 1960s. An important element in the context for its popular vogue was the disillusionment of many French intellectuals with politics. The Communist Party's prestige as organizer of the war-time Resistance had been diminished by its failure to seize state power, and the success of the British and Americans in forestalling communist electoral victories seemed to ensure that no Marxist transformation of French society was in the cards. The long, losing, and bitterly divisive struggle of the succession of national governments leading to De Gaulle to hold on to the overseas empire in Vietnam and Algeria further discredited the state, and with it, for many, politics *tout court*. Neither the mechanistic determination of Marxist historical materialism nor the authoritarian voluntarism of the Stalinist Third International seemed to offer a viable way out. The discrediting of the democratic left by the collapse of the Popular Front and the defeat of the Republic in the war, and the opprobrium attaching to the collaborationist right, rendered the other options of traditional French intellectual politics equally unappealing.

Politics, both politically and intellectually speaking, thus seemed to many a dead end. To many of those so disillusioned, the subjective voluntarism of Sartre's existentialism appeared as a pointless indulgence in nostalgic illusions of the possibility of affecting history through individual action.

In this context of political defeat and disillusion with the possibilities of subjective action, structuralism, with its discovery of a Platonic world of mental phenomena conceived on the model of Saussurrean *langue*, immune from material determination, historical forces, or the effects of social activity, and equally insulated from the illusions of subjectivity, transmuted the political alienation of a generation into the appearance of an apolitical, scientific approach capable of penetrating levels of human psychological and cultural reality inaccessible to either traditional Marxism or Sartrian phenomenology. Avoiding both the Scylla of historical-materialist determinism and the Charybdis of vanguard-party subjective voluntarism, it successfully seized intellectual hegemony from both traditional Communist Party Marxism and Sartrian phenomenology, and held it securely well into the 1960s.

Until 1968, in fact. The Events of May 1968 struck a mortal blow to the hegemony of structuralism in its received Lévi-Straussian and early Foucauldian manifestations, as well as to the remaining claims of Marxism (in both its orthodox Third International and Althusserian structuralist forms) and Sartrian phenomenology, to hegemony in the hierarchical order of French intellectual discourse. The blow fell not because the 1968 events failed as a political uprising, but simply because they happened at all, on a massive scale that made them undiscountable, even by a structuralism theoretically predisposed to discount the importance of events. Neither Marxists nor Sartrians nor structuralists could come up with a convincing explanation for what was happening, much less offer effective leadership. Their claims to privileged insight into historical dynamics and the deep mental and ideological structures supposedly ordering human consciousness and action were thus alike discredited.

What made 1968 so hard to understand in terms of any of the hegemonic theoretical perspectives was that it was not primarily an uprising of any of the familiar French types (a class struggle of workers against capitalists, a liberal movement against political autocracy, or a rightist putsch against the parliamentary republic) but a rebellion of students, young professionals, and aspirants to places in the new "middle-class" culture of capitalist consumerism against the rigid, inefficient and restrictive institutional structure of the national culture. Elements of the working class participated, but this remained of secondary importance and did not determine the character and course of the May events (and was far more firmly suppressed by French security forces). As Debray put it,

May was an individualist revolution, necessary for the elimination of "nation" and "proletariat" as the two main remaining obstacles to "the free flowering of capitalist ideology ... May was the victory of the individual subject ... over the collective subject (nations, classes)." (Debray 1978, quoted in Ferry and Renaut 1990: 45)

The revolt began against the institutions of higher education but quickly extended to the whole hierarchical organization of French bureaucracy, society and intellectual life. Its focus was on repudiation of the hegemonic notions of "system," "structure," and "totality" (epitomized by structuralism and Marxism) which seemed so consonant with the established social, cultural, political, and bureaucratic "system" against which the rebellion was directed.

In this latter respect, if only in this, the 1968 uprising was a smashing success. Perhaps the slogan most frequently shouted and scrawled on the walls of the *Quartier Latin* by the participants in the May events, and the one that best epitomized the character of the rebellion, was *"Prenez la Parole!"* ("Speak Out," literally "Seize the Word!," i.e., take it away from those who have monopolized it): *Parole*, as in the opposite of *Langue*, the site of deep structures and the privileged sanctuary of structuralist hegemony. *Parole* that had to be seized from the established academics and intellectuals who had monopolized and controlled intellectual discourse in the tightly hierarchized world of French academic and intellectual life. The Events of 1968 were about speaking up and acting up, escaping from the confines of orthodox cultural as well as political "structures." "Resistance" thus assumed the prototypical form of public "discourse" rather than covert sniping from the *maquis*.

As a movement for egoistic, individual freedom, 1968 rapidly came to include among its causes demands for total personal liberation, for bodily and erotic unrepression, for escape from all bourgeois social disciplines. Marcuse's ideas on the connection of erotic expression and political resistance had enormous influence: his *Eros and Civilization* was one of the main texts read and cited by participants in the Events. The body moved onto the revolutionary agenda as the subjective, erotic, affective antithesis to the austere and abstract intellectual structures of Lévi-Strauss, Althusser and Sartre.

One of the main reasons that these reigning savants had such trouble figuring out what was happening was that they found it almost impossible to recognize that the revolution was in no small part against *them*. Their manifest incomprehension and ineffectiveness in the event sealed the triumph of the *Soixante-huitards* on the intellectual front. It was, of course, a success wrapped in a failure. Although for a few days the crowds battling the police seemed close to bringing down De Gaulle, in the end their failure to overthrow any part of the received institutional order or to produce an organized political opposition that could survive the Events themselves only reinforced the dismal calculus of political futility which had provided the main condition for the rise of structuralism in the first place. Hence the contradiction: a successful intellectual revolt against structuralism ended by reinforcing the historical-political conditions of structuralist hegemony.

In the scramble to re-establish intellectual hegemony that followed the earthquake of the 1968 Events, it was clear that preeminence would go to the thinker who could most adroitly embody and appear to reconcile these contradictory features of the new intellectual landscape. As a general theoretical framework, structuralism continued to hold a decisive advantage, as its fundamentally apolitical and ahistorical idealism of linguistic and psychological structures was the most consistent with the failure of the Events to produce significant historical or political change. The problem, from the structuralist point of view, was how to encompass and co-opt the activist, politicized, intensely subjective, affectively and erotically charged, defiantly material discourse and acting out of the *Soixante-huitards* into structuralism's quietist, apolitical, anti-subjective, affectless, framework of linguistic idealism, with its relegation of subjectivity, speech and action to the status of epiphenomena of the abstract, objective mental structures of *langue*. Since the 1968 revolt had been explicitly directed against all of these aspects of structuralism, the overriding problem was thus how to continue to be structuralist at the level of general assumptions while appearing to be anti-structuralist on all specific points. This comes to more or less the same thing as integrating matter within anti-matter; but as events were to prove, what is impossible in physics may be achieved in the hyper-reality of French social philosophy.

Foucault's "poststructuralist" resynthesis of his earlier relatively orthodox structuralist position, which appeared complete in its essential outlines in the first volume of *The History of Sexuality* eight years after the 1968 Events, can be understood as a successful attempt to do precisely this. Foucault's new position was not so much "post"-structuralist as a continuation of structuralism by other means. Its essential moves were to retain the essential linguistic idealism of structuralism but to transfer the linguistic focus from *langue* to *parole*, retaining Saussurre's conception of the two categories as the complementary structured and unstructured aspects of language (respectively). Where orthodox structuralism had conceived *langue* as the manifestation of an effectively transcendental structure of the mind (*l'esprit humaine*), Foucault made *parole*, i.e., "discourses," the manifestation of an equally transcendental, extra-historical demiurge called "Power." The locus of the abstract, cognitive structures of the old structuralism of *langue* had been the mind; the locus of the concrete discourses of the new poststructuralism of *parole* (the site of their operations and effects, and the medium of their self-realization) could therefore, by contrast, only be the body.

Thus was born the poststructuralist body, the "site of all controls," i.e., the context of the operations (or "discourses") of "power" as for structuralism the mind had been the context for the structures (*langue*) of thought. As the deep structure of the mind had been held in orthodox structuralism to generate specific cultural structures (myths, totemisms, languages, epistemes), so in Foucault's poststructuralism, power is held to generate specific discourses of control and discipline, which include, as their predicates, the objects of these disciplines and controls: bodies and their desires.

In a revealing passage, Foucault contrasts his new theoretical conception of the body with Marxian materialism on the one hand and (implicitly) his previous structuralist concerns with "knowledge" and *epistemes*:

Question: Would you distinguish your interest in the body from that of other contemporary interpretations?
Foucault: As regards Marxism, I'm not one of those who try to elicit the effects of power at the level of ideology. Indeed I wonder whether, before one poses the question of ideology, it wouldn't be more materialist to study first the question of the body and the effects of power on it. Because what troubles me with these analyses which prioritise ideology is that there is always presupposed a human subject on the lines ... [of] classical philosophy, endowed with a consciousness which power is then thought to seize on. (Foucault 1979: 58)

Foucault's claim is that the body is a "more materialist" substitute for the epistemological subject of social consciousness, and the study of "the effects of power upon" the passive physical body is similarly more materialist than the concept of ideology, i.e., critical analysis of the way social discourses misrepresent the material realities of social production, exploitation, and power. Since Foucault conceives the body essentially as an object, his claim that it is "materialist" can only mean that his conception of "materialism" is of the Feuerbachian or eighteenth-century French varieties that Marx rightly criticized as sharing the "contemplative" orientation of "classical (pre-Hegelian) philosophy." Foucault seems unaware of Marx's substitution of "practico-sensuous activity," a simultaneously subjective and objective continuum of mutually transformative interaction, for the classical subject–object dichotomy as the basis of his "materialism." It is Foucault, not Marx, who claims to found a materialist position on a "classical" philosophical notion of the body as an inert, subjectless physical object.

Even by this vulgar standard, however, Foucault's claim that his poststructuralism is "more materialist" than Marxism because of its grounding in that "material" object, the body, is transparently spurious. Foucault's body has no flesh; it is begotten out of discourse by power (itself an immaterial, *mana*-like force), and the desires that comprise its illusory subjectivity are themselves the predicates of external discourses of power rather than the products or metaphorical expressions of any internal life of its own. It is thus no less an idealist figment than the conceptual body of a Lévi-Straussian totem animal.

Foucault's transformation from "quasi-structuralist" to "poststructuralist" (Merquior 1986: 142), and the specific form of his poststructuralist synthesis with its foundational categories of body, power and discourse, were constructed primarily through Foucault's reworking of ideas drawn from his other major intellectual antecedent (apart from structuralism): Nietzsche. Foucault's relation to Nietzsche, both before and after his conversion from "quasi-structuralism" to poststructuralism, is complex and various, and has been the subject of extensive commentary by Foucault scholars and critics (e.g., Ferry and Renaut 1990; Merquior 1985, 1986; Sheridan 1980; Stauth and Turner 1988; Turner 1984, 1993). Foucault's use of Nietzsche, like his use of structuralist ideas, is partial and selective, and what he omits of the Nietzschean heritage is often as significant as what he makes use of. Certainly Nietzsche's conception of the body as the basis of consciousness, knowledge, and culture, and his substitution of bodily drives, interpreted through metaphor, for "truth" and knowledge of objective reality may be seen as foundational in relation to Foucault's poststructuralist synthesis. Nietzsche's notion that naming things determines their essence likewise foreshadows Foucault's notion of discourse. The implicit contradiction between Nietzsche's conceptions of the physical body and its drives as the prior source of what we can interpret of the world and his assertion of the power of language (naming) to determine essence is replicated in the contradiction between Foucault's notion of the body as the product of discourse and the body as the pre-discursive source of pleasure and "resistance." Yet Foucault fails to follow up some of Nietzsche's most interesting insights into bodiliness, for example, Nietzsche's notions of the concrete plurality of drives and thus of phenomenal "bodies" (contrast this with Foucault's treatment of the body as abstract conceptual unity), and his development of his notions of interpretation and metaphor as the subjective processes through which the reality of the body is translated into cultural forms of consciousness (Blondel 1991: 203 ff.). Foucault's failure to develop this aspect of Nietzsche's thinking on the body may perhaps be due to its association in Nietzsche with notions of subjective will and activity. Bryan Turner has also noted Foucault's failure to take account of Nietzsche's crucial insistence on the importance of the body as potential (B. Turner 1984: 248).

Foucault's theoretical body, in fact, embodies the full range of contradictions inherent in the intellectual politics out of which poststructuralism was produced. Although he presents his post-1968 synthesis of body/discourse/power as an alternative to the individualism and transcendental subjectivism of bourgeois liberal and classical philosophical (Cartesian, Kantian) approaches, Foucault ends by replicating their premises in essential respects. As a featureless *tabula rasa* awaiting the animating disciplines of discourse, Foucault's body possesses an a priori individual unity disarmingly reminiscent of its arch-rival, the transcendental subject. As objects of the disciplines of medicine, psychology, sexology and other discourses of power, bodies are inherently individual units. Although he develops his own position ostensibly as a critique of these discourses, moreover, Foucault accepts their fundamental premise in this most essential respect: for Foucault, the body is once and for all an individual body, bounded by its skin and congruent with an individual social person in the modern West. When Foucault discusses populations consisting of multiple bodies, he assumes the disciplines and discourses are addressed uniformly to all the bodies comprising the population. His bodies are at once abstract and uniform, infinitely malleable objects of manipulation by power, undifferentiated either internally into specific organic parts or externally into classes or even, with startlingly few exceptions, genders. As social theory, the result is indistinguishable in form from Hobbesian or liberal-bourgeois individualism.

"The body," for Foucault, however, has a contradictory dual existence: on the one hand, as the product of discourses, it is a historically contingent creation of power; but as that which is disciplined and molded by discourse, that entity whose capacity for pleasure is transformed by power into manipulable desires, the body exists in an ineffable state as pure potential for pleasure prior to all social effects and beyond the limits of Foucault's analysis: in effect, a transcendental object. At times, also, it seems, it is even a transcendental subject. At certain points, for instance, Foucault treats the body as not only created by power, but also as "resisting" power, not, to be sure, in collective political alliance with other bodies, but through private acts of "deviance" or "perversion" in which its innate capacity for pleasures finds expression in ways at variance from the normative forms of desire enjoined by the disciplines of power (B. Turner 1993: 53–4). "Resistance" is thus explained as a sort of natural (i.e., pre-social and apolitical) emanation of the body, as "power" is conceived as a natural (trans-historical and trans-cultural) emanation of society. Neither has a definable political purpose or specific social or institutional source. In being thus depoliticized and desocialized, Foucault's "resistance" thus ironically becomes, in effect, a category of transcendental subjectivity situated in the body.

The body acts as transcendental subject, or perhaps demiurge, at another massively contradictory moment in Foucault's attempt to historicize his analysis. Foucault divides his discussion of the body into two levels: individual bodies and populations. In a given historical period, the "body" appropriate to that period is created by the discourses of power peculiar to the period (power itself is treated as a universal historical constant, in effect another transcendental category with alienated agentive or subjective qualities). But what creates these specific discourses? Foucault analyzes the characteristic discourses of power of the modern period (e.g., those of medical and sexual science, and institutional forms like hospitals, prisons, and insane asylums) as arising in the eighteenth century in response to the need to regiment the rapidly growing population of Europe, particularly in urban areas. Population growth, or the body in its collective aspect, in short, causes specific discourses, which in turn cause specific bodies – yet the reasons for the population growth remain outside the analysis, leaving it in effect a *deus ex machina*, or in other words yet another effect of the body as transcendental historical agent. Foucault thus has his cake and eats it too.

Perhaps the most glaring contradiction of Foucault's impressive array, and I cannot help but think the most cynically and deliberately deceptive, is his dismissal of the whole concept of ideological criticism – and its goal of a liberating revelation of the reality underlying "discourses" of power – by Marxism and by implication all other strands of social science, philosophy or cultural criticism, followed by his own use of a crude imitation of critical analysis to discredit the possibility of a genuinely political or emancipatory criticism, let alone political action, of "resistance." Foucault's rejection of the Marxian critical concept of "ideology" on the spurious ground that it presupposes a "transcendental subject" has been noted. His own claim that discourses and acts of resistance are themselves *really* forms of the "power" which they imagine themselves to be resisting, however, has precisely the form of the critical arguments he rejects from others (the disclosure of a contrary meaning or reality beneath the surface appearance of some form of social consciousness), except that with Foucault it is deployed *against* the possibility of emancipation rather than in support of it. It is hardly neces-sary to add that Foucault's claims on this point are made with the same absence of supportive analysis or coherent argument that characterize his other main theoretical assertions.

These contradictions in Foucault's "poststructuralist" theoretical writings are not accidental or unconnected, nor are they merely peripheral to his argument. On the contrary, they are central components of the political and intellectual project which that argument was constructed to implement: the co-optation and neutralization of May 1968, and the threat its combination of hyper-liberalism and anarchistic personal expressiveness presented to the hegemonic theoretical, political and ideological perspective of structuralism with which Foucault had identified in his earlier work. This co-optation was to be achieved within a continuation of the essential framework of structuralist theoretical assumptions, and above all the struc-turalist posture of accommodation to and acquiescence in a political *status quo* that appears to render political resistance, and all social action beyond the level of private erotic self-pleasuring, futile.

The connecting thread running through all the contradictions is the attempt to adopt the appearance of politicized, critical thought while simul-taneously denying the possibility of politics or criticism. At the level of "the body," this takes the form of insisting (against the Marxists) that the body is the truly "material" base of social existence and consciousness, while proclaiming (with the structuralists) that the body is an (idealist) construct of discourse. Again, at some points, Foucault appears to give the body an overtly political role as the (implicit) subject of resistance, while at others he denies the possibility of political resistance by insisting that it is merely an effect of power (assertions of the latter sort far outweigh the former in his work).

It is frequently said that despite regrettable a prioris such as those I have just catalogued, Foucault has provided original and valuable insights that have opened new vistas for intellectual and social history, literary theory, cultural criticism, anthropology, and feminist theory. These insights are generally held to include the following: that the reality of the body and its desires are historically determined, that this determination is essentially political, consisting in the operations of power and the resistance these operations arouse; that the body is the privileged or even the sole real object and site of these political operations, and therefore, in effect, the real or primary matter of politics and historical determination as such; and that since the medium of these political and historical operations is discourse, to analyze discourses of power about the body and the disciplinary practices these discourses enjoin is to engage in political action, perhaps even resistance, of the most meaningful, or perhaps even the only genuine and possible kind.

My assessment of Foucault's contributions and the influence he has exercised, and continues to exercise in humanistic and social studies, is at some variance with such evaluations. Leave to one side that Foucault is hardly the first or the only thinker to have suggested that the social, cultural, psychological and personal realities of the body, together with bodily functions, products, desires and disorders are historically and culturally variable, or that such variations are part and parcel of formulations of gender roles, sexuality, race, diet, hygiene, sickness etc. that may in turn be connected to relations of inequality, domination and social control. As I have probably by now made overabundantly clear, I regard the ostensibly "political" character of Foucault's thought as a deceptive rhetorical posture that conceals Foucault's essential message of the futility of political action and the impossibility of meaningful emancipation from politically imposed controls. It is important to insist that this depoliticization holds even at the level of the body's individual erotic practices, which Foucault defines as the only possible "resistance" on condition that they be limited to nothing more than anarchic private acts of "deviance." For Foucault, in other words, the personal is not the political (but then it's not personal either).

Perhaps more serious in its implications for humanistic and anthropological studies, however, are Foucault's gratuitously inadequate and one-sided conceptualizations of his major theoretical categories: power, discourse, and the body itself. It is difficult for anyone who attempts to make serious analytical use of Foucault's categories to prevent these shortcomings, and the tendentious and contradictory asumptions from which they derive, from affecting her or his own analysis.

To begin with "discourse." In attempting to carry over the linguistic basis of structuralist theory in the form of his category of "discourse," Foucault contradictorily attempts simultaneously to retain and collapse Saussure's fundamental distinction between *langue*, the synchronic, abstractly conceptual aspect of language that served as the paradigm for structuralist theory, and *parole*, the complementary diachronic aspect of language as material acts of speech that corresponds to Foucault's "discourse." For Saussure, *parole* or discourse is speech by an individual subject, who purposefully combines elements of *langue* to form a specific message, which involves, among other things, the use of sign-elements to refer to objects in the real world, notably including the context of speaking. "Discourse," in sum, is agentive, subjective, intentional, meaningful, and intrinsically referential and context-relative. These features are retained by contemporary linguistic discourse analysis (which no longer accepts Saussure's distinction of *langue* and *parole* in the terms in which he formulated it).

In spite of its ostensible identification with material acts of speaking (*parole*), Foucault's "discourse" has none of these distinctive features of discourse as understood in linguistics. As the sociologist Bryan Turner has pointed out in his semi-Foucauldian book, *The Body and Society*, Foucault's notion of discourse is mechanical and in effect self-determined by its own rules, independently of speakers. It precludes any role for agency, and is presented as operating independently of the social groups that are its primary carriers, forgetting that discourses are social products that express the intentions and perspectives of specific social actors. Its messages are abstracted from the contexts of speech. There is no consideration of discourse structure or organization, the logic of discourse being directly equated with its social effects (B. Turner 1984: 174–6).

What all this means is that Foucault's "discourse" is really not discourse at all, in the sense of social speech or *parole*, but a sort of *langue* without the structure, that retains the specific structuralist (if not Saussurean) idea of *langue* as a set of decontextualized, subject-independent, self-acting and self-imposing patterns of extra-individual signification. Foucault's "discourses" and "disciplines" of "power" are in effect just that: self-acting and self-imposing significations, with no element of intention (i.e., interpretable meaning). Beyond sloughing off the now untrendy dimension of structure, the only real change is terminological: the substitution of the new and decidedly trendy term, "discourse," for "structure."

This brings us to the second of the three categories: "power." Foucault's vague and undefined use of this term serves mainly to separate power and its exercise from any specific social agents or groups and to fetishize it as a sort of social *mana* that emanates "from below," that is, from all aspects of social relations rather than those "from above," i.e., from political leaders or dominant classes (this may be Foucault's closest approach to a Sartrian thought: *l'enfer, c'est les autres*). "The body" as "the site of all controls" plays a key role in this eminently depoliticizing move of separating power, and oppression, from the oppressors who wield power. Of course it is not true; the body is not the site of the vast majority of social controls and exercises of power, particularly in the modern societies with which Foucault is concerned. Most effective social control in such societies is exercised at "sites" or levels of social organization well above that of individual bodies or even whole populations considered as aggregations of bodies. Examples are legion: capitalist relations of production, which since the times of primitive accumulation in the central capitalist countries have depended on the ownership of inanimate means of production rather than the direct coercion of workers' bodies; national legislatures that are effectively controlled through monetary contributions and hegemonic influence of representatives of the ruling class; and university administrations which routinely coerce professors to teach unpopular courses despite the unavailability of torture.

Part of the problem is that Foucault fails to differentiate among the qualitatively different types of effectiveness in social relations: "power" is one mode of effectiveness, but influence is another, appeals to shared values or commitments another, and seduction yet another. All have their respective discourses, and their specific modes of resistance. This is another way of saying that societies are never simply systems of repressive control maintained by power, but involve a plurality of modes of obtaining consent and conformity, as also of forms of resistance. To the extent that these different discourses impinge on the bodies of members of the society, they impinge in different ways and with different meanings and effects. It is also of course essential to remember that not every "discourse" is specifically political or involves power. Serious social or cultural analysis requires clear criteria for differentiating what are discourses of power from what are not, and what other modes of discourse may be socially recognized and employed. There is no empirical or theoretical warrant for defining all socially or even politically significant discourse as "discourses of power" on a priori grounds, or for maintaining on principle that all discourses of the body must be interpreted as discourses of power and therefore as essentially about political discipline and control. To accept Foucault's arguments to this effect is to commit oneself to tendentiously Procrustean interpretations of the meanings of the social, political and bodily discourses of any society, ancient or modern, classless or capitalist.

To focus on the specific theoretical shortcomings of Foucault's category of "power," however, amounts to tacitly consenting to confine critical discussion within the theoretical framework that term purports to define. This is already to miss the main point, which is that the primary function of Foucault's deliberately vague and abstract discourse of "power," so devoid of specific content, is not to serve as an alternative theory of the social or historical determination of bodiliness or forms of subjectivity, but to substitute for such a theory, providing its appearance but avoiding its substance. It is a rhetorical trope in a highly subjective discourse of personal power, a weapon of attack and competitive disempowerment of potential critics and rival theories, not a tool of analysis or understanding. Itself lacking any definable theoretical content, and thus relatively invulnerable to specific theoretical criticism, it serves as the ideal general pretext for dismissing all other social theories as so many self-deluded instances of discourses of power. By the same token, its rhetorical self-presentation as a disclosure of the universality of the deceptive manipulation of discourses of rational understanding, subjective identity, and resistance by power reflexively serves to suggest that it itself is the sole exception to the general phenomenon it purports to identify: that it is, in sum, the one discourse about power that is not itself a discourse *of* power. The earnest attempts of devoted Foucauldians to get Foucault to be a little more specific about his notion of power thus missed the point (with what predictable lack of result can be seen in, e.g., Dreyfus and Rabinow 1983: 208–28). What they (correctly enough) saw as the weaknesses of Foucault's theoretical discourse about "power" were, for Foucault's own purposes, precisely its strengths *as a discourse of power*.

To move on, finally, to the third fundamental category: the body. Bryan Turner (1984: 250) makes the point that Foucault "fails to provide an adequate phenomenology of the body [because he] abandons the idea of the body as sensuous potentiality . . . [and fails to recognize that] embodiment is more than conceptual; it is also potentiality." As Turner (ibid.: 248) observes, this appears to leave no room for Nietzsche's essential notion of a "life-affirming instinct," through which Nietzsche preserved a connection between bodiliness and subjective activity (e.g., the will to power). It thus provides no basis for the really important bodily-identified political movements of Foucault's (our) time: gay liberation and the feminist movements for gender equality and reproductive rights. One might add to this the more fundamental point that what Foucault's notion of the body as an essential inert or passive conceptual category excludes is the body in its socially fundamental aspect as material activity, including the entire range of concrete bodily processes and faculties that are directed and effected, with varying degrees of success, within the schemas of acting, perceiving, cognizing and feeling that make up social representations of the body, and which comprise the material content of both moment-to-moment interactions of the embodied actor with the object world and longer-term processes of socialization.

Bodies and anti-bodies

My concern in this extended critique of Foucault on the body has not been simply with Foucault. I have focused on his work because his treatment of bodiliness is the most fully developed and influential example of a much wider tendency in social theory that spans many academic disciplines, theoretical schools, and now also national traditions of social theory, particularly French, British, and American. The importance of Foucault's work as paradigm and prototype of strong theoretical and ideological currents in contemporary thought is amply attested by the widespread fashionability of his terms, categories, and agenda, even among researchers and writers who have never read his work. Whether they assert a central role for the body as a component of social, cultural, psychological, or phenomenological reality or whether they deny physical bodiliness any role in these same modes of reality, on grounds of an idealist anti-materialism or anti-empiricism, all of these positions converge, remarkably on the point of treating bodies, bodily functions, and powers as products or projections of cultural discourses or symbols rather than as pragmatic individual and social activities of production and appropriation. There is clearly an antipathy of much contemporary social theory to flesh and all that goes with it, particularly the carnal aspects of reproduction, involving as it does the biological interdependence of bodies (male and female, infantile and adult). It is above all in this connection that one must confront the plural aspect of the body as a relation (both physiological and social) among bodies, rather than the singular and individual aspect of the body as the subject of sensations of erotic pleasure or pain. To emphasize the latter aspects of bodiliness to the virtual exclusion of the former is to distort the nature of the body by suppressing its collective, plural, and social aspect, leaving only its private, individualistic aspect, and thus to end by reinforcing, rather than criticizing, the individualist bias of received philosophical and utilitarian liberal political perspectives. Rather than theorists of "the body," it therefore seems to me that Foucault and other contemporary theorists might more justly be called "anti-bodies."

Anti-bodies are defensive organisms, activated to repel unwelcome intrusions of alien realities upon self-contained bodies of flesh or theory. The Events of May 1968 represented such an unwelcome encroachment upon the *corps* of aging thinkers and theories which constituted the French intellectual and academic establishment. The intrusive reality they embodied, however, did not go away when the last *pavé* had been thrown in 1968. The themes of personal, sexual, and bodily emancipation have remained linked through a variety of cultural manifestations, social movements, and political alliances. They have also steadily grown in assertiveness, until they have become permanent forces to be reckoned with in all the advanced capitalist societies.

To return to the point made earlier: what is essentially at stake in the contemporary concern to theorize the body and the struggle to liberate or control it is control of the relations of personal production, meaning both the production of personal identity and the material conditions of personal bodily existence. In this respect, the women's movement for gender emancipation and reproductive rights, gay and lesbian activists, and other exponents of the politics of identity such as human and indigenous rights advocates, and important segments of the black and environmental movements, represent an unsettling threat to the *status quo* represented by the political rulers and dominant elites of the received social and cultural order, as well as those who control the relations of commodity production, and that part of the intellectual establishment which has invested in theorizing, justifying and defending the received order as the condition of its own hegemony within the academic and cultural establishment.

If the new movements succeed in asserting control of their personal relations of production, might they not be led to widen their demands for control of other conditions of existence, even other relations of production? If they succeed in crumbling the cake of custom and dismantling the customary forms of hierarchy that have defined them as unequal and under the control of others, what will become of the hegemonic political and intellectual arguments that insist that the received forms of hierarchy and value consensus, or in other terms the "discourses of power," must be accepted as the natural, or at any rate the inevitable order of society? And if these legitimating political and theoretical arguments are successfully defied, what will become of the hegemonic position of those who make them? The threat posed by these new movements, however, is double-edged, as much to themselves as to traditional social mores and hierarchies. Here they face the danger of subverting and neutralizing their own potential for political and cultural transformation through their own ideological mystification of the social roots of, and political-economic limitations on, the overtly individual forms of bodily and personal liberation they advocate.

In this perspective, the poststructuralist attempt to substitute a desocialized, depoliticized, and desubjectified theory of the body for a conception of embodied subjectivity and material bodily activity through which the body participates in multiple bodies and interpersonal social relations may stand as the prototype of the reaction of established political and intellectual elites to the new politics of personal identity and bodily emancipation. At the same time, however, it represents the prototype of the ideological distortions of the new forms of bodily consciousness that pose perhaps the greatest danger to the potential of the new movements for political effectiveness and social transformation. The conflict between these reactionary and emancipatory aspects of postmodern culture is epitomized by their opposing views of the body. For poststructuralism, the body is "the body": an abstract, singular, intrinsically self-existing and socially unconnected, individual; the social behavior, personal identity and cultural meaning of this entity are passively determined by (disembodied) authoritative discourses of power. For the new political movements of personal-social-cultural-environmental resistance, by contrast, "the body" consists essentially in processes of self-productive activity, at once subjective and objective, meaningful and material, personal and social, an agent that produces discourses as well as receiving them.

The body, in sum, has become the focus of fundamental contradictions between the emancipatory and repressive tendencies of the social, cultural and political order of contemporary capitalism. But we are no longer really speaking here of the same "body," a self-evidently singular corporeal entity mediating both sides of the contradictions in question. Two very different bodies, the opposites of each other in *every* significant respect (in the terms I have suggested, authentic social bodies and poststructuralist anti-bodies) confront each other across the battle lines of identity politics and the theoretical divisions between the social thinkers associated with the opposing sides. The opposition between these two bodies involves theoretical, ontological and epistemological issues that are central to contemporary social thought.

ACKNOWLEDGEMENTS

Jane Fajans read and criticized successive drafts of this paper. I have benefited from many conversations with Steve Sangren about Foucault and poststructuralism generally, and am obliged to him for numerous ideas and bibliographical sources. Tom Csordas and an anonymous press reader also provided useful comments, which have been taken into account in the final draft. Needless, to say, however, the responsibility for the ideas set forth here is wholly my own.

REFERENCES

Blondel, Eric (1991) *Nietzsche, The Body, and Culture*. Stanford, Stanford University Press.

Bourdieu, Pierre (1984) *Distinction: A Social Critique of the Judgement of Taste*. London: Routledge and Kegan Paul.

Debray, Regis (1978) *Modeste contribution aux ceremonies officielles du dixieme anniversaire*. Paris. Masepro.

Dreyfus, Hubert L. and Paul Rabinow (1983) *Michel Foucault: Beyond Structuralism and Hermeneutics*. Chicago: University of Chicago Press.

Ehrenreich, Barbara, and John Ehrenreich (1979) The Professional-Managerial Class. In Pat Walker, ed., *Between Labor and Capital*. Boston: South End Press.

Ferry, Luc and Alain Renaut (1990) *French Philosophy of the Sixties: An Essay on Antihumanism*. Amherst: University of Massachusetts.

Foucault, M. (1972) *Power/Knowledge: Selected Interviews and other Writings, 1972–1977*. New York: Pantheon.

(1979) *The History of Sexuality. Volume I: An Introduction*. London: Tavistock.

Fraser, N. and L. Nicholson (1989) Social Criticism without Philosophy: An Encounter between Feminism and Postmodernism. *Theory, Culture and Society* 5 (2–3): 373–94.

Giddens, Anthony (1991) *Modernity and Self-Identity*, Stanford: Stanford University Press.

Marx, Karl (1964) Estranged Labor. In K. Marx, *Economic and Philosophic Manuscripts of 1844*, Dirk, J. Struik, ed. New York: International Publishers.

Merquior, J. G. (1986) *From Prague to Paris: A Critique of Structuralist and Post-Structuralist Thought*. London: Verso.

(1985) *Foucault*. London: Fontana.

Pfeil, Fred (1990) *Another Tale to Tell: Politics and Narrative in Postmodern Culture*. London: Verso.

Sheridan, A. (1980) *Michel Foucault, The Will to Truth*. London: Tavistock.

Stauth, G. and Bryan S. Turner (1988) *Nietzsche's Dance: Resentment, Reciprocity and Resistance in Social Life*. Oxford: Basil Blackwell.

Turner, Bryan S. (1984) *The Body and Society: Explorations in Social Theory*. Oxford: Blackwell.

(1993) *Regulating Bodies: Essays in Medical Sociology*. London: Routledge.

2   Society's body: emotion and the "somatization" of social theory

M. L. Lyon and J. M. Barbalet

This chapter shows how the body may be incorporated into theory simultaneously at the individual and social levels, and in the biological and cognitive domains. We will show that it is through emotion (feeling/sentiment/affect) that the links between the body and the social world can be clearly drawn.

We begin with a brief consideration of two recent sources of the ideological representation of the body in modern Western society: the "consumerist" body, and the body as a terrain of medical theory and practice. The conceptions of the body implicit in these converge in so far as they postulate the body as more or less passive in the face of subordinating forces. These forces render it partial in terms of either its full potential capacity or its relationship with the human person embodied in it. These considerations are of concern because scholarly treatments of the body are developed in contexts unavoidably implicated in ideological constructions.

A quite different conceptualization of the body is proposed in the second section of the chapter. A model is developed whereby the body is understood not merely as subject to external agency, but as simultaneously an agent in its own world construction. Here the body is not treated as a discrete physicality external to the self. The body is intercommunicative and active; and it is so through emotion. Emotion is an integral part of all human existence. Emotion activates distinct dispositions, postures and movements which are not only attitudinal but also physical, involving the way in which individual bodies together with others articulate a common purpose, design, or order. The role of emotion in this is essential, for we will show that emotion is precisely the experience of embodied sociality.

This model of the body is supported in the third section of the chapter by a brief reading of the ethological and phenomenological literatures. These articulate a basis for an understanding of the integral role of emotion in social life in a manner which clearly indicates both an embodied form of emotion, and also the role of emotion in bodily agency.

48

Contemporary ideological constructions

Since the mid-1980s there has been a growing concern with the human body in social theory. The treatment and understanding of the matter has been almost without exception one-sided, for it has focused on the body as an *outcome* of social processes. This is the strength of the work, for example, of Michel Foucault (1979, 1980) and Norbert Elias (1978 [1939]). They each indicate the ways in which human bodies have been transformed by social processes. But Foucault and Elias have little to say about bodies as social agents. The bodies they deal with are the bodies of individuals subjected to forces over which they have no control.

For Foucault, the body is the text upon which the power of society is inscribed. In studies of particular organizations – hospitals, asylums, prisons, schools – Foucault has developed the idea that a new mode of power emerged in the eighteenth century which does not relate to subject populations as an external and repressive sovereign power, as it had up until this time, but functions rather as a disciplinary power which is "a synaptic regime of power, a regime of its exercise within the social body, rather than from above it." He says that "power reaches into the very grain of individuals, touches their bodies and inserts itself into their actions and attitudes, their discourses, learning processes and everyday lives" (1980: 39).

Like Foucault, Norbert Elias (1978 [1939]) develops an account of the way in which bodies are changed through social processes. But unlike Foucault, in his study, *The Civilizing Process*, Elias provides a central role for emotion in the socialization of natural and bodily functions. Elias details the variation in attitude through historical time toward certain "natural functions:" eating, nose blowing, spitting, bedroom behavior and relations between the sexes. In doing so, Elias shows how the body, or more particularly bodily functions, are shaped through historically driven social forces. Elias's argument is that in the course of its development "civilization" has entailed changes in human behavior which include the restraint, regulation and transformation of bodily and physical impulses. In particular, the role played in the regulation of behavior by changes in the feelings of shame, embarrassment and delicacy are highlighted by Elias. Feelings of shame and embarrassment constitute a compulsion to check one's own behavior, and to enforce a conformity on oneself with what the subject feels to be appropriate standards. Thus, as Elias puts it, the "feeling of shame is clearly a social function molded according to the social structure" (ibid.: 138).

Elias argues that the feeling of shame is a social restraint and that it is only in modern society that feelings of shame are generalized to the transgression itself irrespective of the social context in which it occurs (ibid.: 134). Elias explains this change in terms of the development or intensification of the social division of labor (ibid.: 137–8, 152). It is under these conditions that "everyone becomes increasingly dependent on everyone else ... [and that] the socially superior feel shame even before their inferiors" (ibid.: 138).

An implication of this is a tendency to regard the body as a more or less passive recipient of social processes. Elias refers, for example, to the "regulation or molding of instinctual urges," to the "transformation of aggressive expression of pleasure, into the passive, more ordered pleasure of spectating," and to the "whole molding of [man's] gestures" (ibid.: 139, 202–3). In Elias, the embodied social *agent* remains underdeveloped.

An adequate understanding of social agency requires a concept of embodied agency. This is because emotion, which is necessarily embodied, functions in social processes as the basis of agency. Emotion has a role in social agency as it significantly guides and prepares the organism for social action through which social relations are generated. The body cannot be seen merely as subject to external forces; the emotions which move the person through bodily processes must be understood as a source of agency: social actors are embodied.

Action is necessarily bodily and involves physical movements of great complexity which are not merely individual but also always associative. There is now a growing awareness of the somatic dimensions of action in social theory. When over thirty years ago the sociologist Dennis Wrong asserted that "we must start with the recognition that in the beginning there is the body" (1961: 191), there were few social theorists who would have agreed. This is no longer so today. But the current interest in the body in social theory generally fails to emphasize the body as agent, and has little to say about the role of emotion in the social body, in spite of Elias's example. The question of why the interest in the body has developed in recent years is therefore subordinate to the question of "what is the body which draws attention?" It is not obvious what "the body" is, as different concerns have different conceptions of it.

The understanding of the body in scholarly writing is not remote from the conventional and everyday notions of it found in popular discourse. Concepts arise in social and historical settings, and scholarly as well as everyday notions of the body share generalized social milieux. It will be shown that conventional conceptions of the body are well articulated with scholarly notions of it.

Whilst the scholarly interest in the body is relatively recent, in general people have always, indeed necessarily, had a practical interest in bodies. What we feed and clothe are our bodies; we seek comfort and pleasure in our own and other's bodies; we are concerned with the health of our bodies; and, in general we use our bodies as instruments of our desires. In addition to these more or less direct applications, which are conducted below the threshold of consciousness, it is with our bodies that we express our feelings and dispositions and actively occupy the spaces we inhabit.

In a real sense the body mediates biological and social processes. The conception of the body which operates at any given time in any particular society will therefore be characteristic of its circumstances. For example, without proposing that these are in any sense exclusive sources for an understanding of the body, the current Western conception of the body is molded by two particular influences which fashion our scholarly but no less socially embedded understanding of it. The first is the "consumerist" body, the second is the body as a terrain of medical practice.

To these two bodies could be added others, of course, as Bryan Turner (1984), John O'Neill (1985) and others demonstrate. But the purpose here is not to generate symbolic "bodies" which can be related to physical bodies analogically or metaphorically. Rather, it is to indicate the main sources of ideological representation of the human body through the grounded experience of concrete practices in which bodies are unavoidably implicated.

The idea of the "consumerist" body is outlined in Mike Featherstone's discussion of "The Body in Consumer Culture" (1982). Featherstone argues that since the 1920s, with the development of the popular cinema, a mass circulation press driven by advertising, and rising real wages and consumer credit, images of youth and beauty became loosely associated with the desire for mass consumption goods of various types. Under these circumstances the body is not only a vehicle of pleasure but a new expression of self. The consumerist body is the body we dress, feed and exercise as projects in their own right. This is the body which exists as the body we, but especially others, see. It is difficult to place too much importance on this fact because in the current configuration of the market economy the body becomes a possession which, through its appearance, provides opportunities for social and professional advancement: appearance, display and the management of impression are the capital goods of the consumer economy (Featherstone 1982: 27–8).

Two things in particular can be added to Featherstone's account. First, it is important to remember that the consumer culture described by Featherstone arose at the time when manual labor, in which the body is active in production, began to decline as a proportion of all paid employment. Developments in productive organization increased the numbers of clerical, supervisory, administrative and managerial jobs. In addition, the growth of commodities markets gave rise to large numbers of sales, service, and accounting jobs. Today the proportion and numbers of manual workers in advanced Western societies is even less than it was in the 1920s, added to which is the fact that in practically all Western economies today the majority of manual jobs are performed by foreign workers.

Thus the laboring body has been displaced from current consciousness and is not merely separated from the consuming body, which exists in its own right, but does not even stand next to it as an alternative conceptualization. The body generated in (post-) modern economic relations is the consumerist body: there are no competing images. With the domination of mental over manual labor the phenomenal experience of the body and its ideological and common-sense representation is consumerist.

The second thing about the consumerist body which needs to be brought out, and which follows from what has just been said, is that it is significantly passive in the way that the laboring body is not. It is passive in that persons are seen to possess their bodies, and it is passive as an object of gaze and exchange. The consumerist body is very much the body we have and do things to. In addition, as it exists in terms of its appearance, the consumerist body is objectified in its relations with others. Even when a person is advantaged in some way by his or her body's appearance, the body is not an agent in any meaningful sense, but is an object of exchange. The significance of this will be indicated below.

The body as a terrain of medical practice, defined and created by medical science, has enormous importance both for our practical or everyday understanding of the body, and for our understanding of approaches to the body in social science. There are three qualities of medical theory and practice which define the body. First, the bio-scientific model at the root of medical epistemology necessarily regards the body as an object external to the enquiries which yield knowledge of it. Second, the medical model contains an implicit authority prerogative in which the practitioner is in control of the body of the patient, and through his or her body, the patient him- or herself. This is linked to the first point above in so far as medical authority is considered to be based on professional scientific knowledge. The skills of medicine are diagnostic and treatment (surgery or drugs) oriented and these require the patient to be subordinate to the practitioner. Third, biomedicine and its practitioners deal with malfunctioning organs or other subsystems of the body, and with the symptoms thereof, not with "the body" as such. It is interesting to reflect why the medical account and not other partial accounts of the body, such as those of morticians, prostitutes, and assassins, for example, have been so important in informing the social-scientific and general discussions of the body.

Following these three qualities of medical theory and practice, the medical conceptualization of the body is clear. The medical body is passive; any active capacities it may display are regarded as internal to its physiology, and these can be revealed to external observers as objective knowledge.

Second, the body is readily subordinated to the authority of medical practice. It is disciplined and made, or at least made better, through the social institution of medicine. Finally, the medical body is a partial body. It is partial in a dual sense: it is the internal body, and it is the body patients have, but not the body patients are in the full sense. This last point is the source of the limitations of the biomedical understanding of healing, made more apparent to medical science itself through recent research developments. This is especially so, for example, in the development of interdisciplinary fields such as psychoneuroimmunology where new research techniques have made possible the exploration of correlations between diverse types of behavioral and physiological data through measurement of immune-system functioning (Lyon 1990, 1993; see also Foss and Rothenberg 1988: 114–31; Haraway 1991: 203–30).

The nature of the "medical body" has been of central importance in the generation of a reaction to it, and an incitement to a new interest in the body (Frank 1990: 131; Scheper-Hughes and Lock 1987). This is particularly true in its relationship to women and "women's complaints." This is to say that there has been a resistance to what has been perceived as the authoritative control of female bodies by a predominantly male medical profession. Reaction to this control (along with resistance to male authority in other domains such as technology, art, literature, sexuality, and consumerism) has been crucial in informing the feminist conception of the body. The idea of "Our Bodies, Our Selves" captures much of the understanding of the body which arises in the feminist reaction to the medical body, namely that "whatever is my body is my self" (Frank 1990: 131). The feminist claim that "what you do to my body you do to me" challenges the ethos of medical practice in so far as it challenges the authority relations of medicine. But feminism can only challenge the epistemological foundations of medicine when it insists that knowledge of the body be taken beyond the realm of cognitive and disembodied practices.

In his review of recent books on the body Arthur Frank reminds us that "[f]eminism as a praxis has taught us to look first for the effects of politics in what is done to bodies" (ibid.: 132). If this were the full extent of the feminist conception of the body it would have to be said that it shared terrain with the medical perspective in so far as the body is regarded as a thing subjected to forces in which it has no formative role. Indeed, as with the work of Foucault, critiques of the passive body may be used to perpetuate that passivity if the fundamental issue of bodily agency in social theory is not addressed. Whether the subjection of the body to some external agency takes the form of consumer imperatives, medical intervention or male violence, the body remains more or less passive in the face of subordinating forces which treat it as or render it partial, either in terms of its full potential capacity or its relationship with the person embodied in it.

Part of the reason for the contemporary scholarly interest in the body is the recentness of consumerism and its critique, the growing power of medical authority, and the rise of the feminist critique. These have shaped the ideological representation of the body. But this representation and its practical everyday reality is not the primary source of academic interest in the body. This interest in the body has a different source. Axel Honneth and Hans Joas (1988: 121) have argued that the critique of the Western "scientific" form of rationality and the quest for "subjectivist" and "individualist" sources of social change, have together introduced the body into current social theorizing. Similarly, Frank cites "the contradictory impulses of modernity" toward "transcendental reason" and toward the fragmentation of life (1990: 132). These are entirely compatible with the idea that within modernism the body can provide a grounded truth. As Frank (1990: 133) puts it: "[t]here is a desire to use the physical reality of body as final arbiter of what is just and unjust, humane and inhumane, progressive and retrogressive". This is particularly the case with respect to the feminist critique of consumerism, medicalization, and science, in so far as "Feminism as a praxis has taught us to look first for the effects of politics in what is done to bodies" (Frank 1990: 132).

The representation of the body in scholarly discussion significantly reflects the ideology of the body found in the consumerist and medicalist formulations. In so far as this is so, the scholarly treatments of the body in society fail to see it as anything but a social artifact, rather than an active source of social processes and institutions. Yet it is primarily because of this latter capacity that the body is important to social theory: an appropriate conceptualization of it forms the basis of a theory of institutions and processes that goes beyond the rationalist and, ironically, individualist accounts of social phenomena.

None of this denies that the body is a subject of (and subject to) social power. But it is not merely a passive recipient of society's mold, and therefore external to it. The human capacity for social agency, to collectively and individually contribute to the making of the social world, comes precisely from the person's lived experience of embodiment. Persons do not simply experience their bodies as external objects of their possession or even as an intermediary environment which surrounds their being. Persons experience themselves simultaneously *in* and *as* their bodies. We all do this especially when we feel the reality of our presence in the world: emotion is central to an understanding of the agency of embodied praxis.

A model of body and emotion in society

In the various representations of the body found in the current literature there is a surprising under-representation of the body as a "corporate" body, not in the strictly legal sense (although that is not disallowed), but in the sociological sense. It is worth recalling that Durkheim, for example, acknowledged a simultaneous individual and social ontology of emotions, as well as their inextricability from thought and belief (Durkheim 1954 [1912]), which is parallel to the argument advanced here.

The body of the Church, for example, exists as a congregation and a community, and, to take a phrase from Lenin, an army can be conceived as "a body of armed men." "Body" in these cases is not a mere metaphor, although it is frequently understood this way, as in the biblical quotation (I Corinthians 12:12): "For as the body is one and hath many members, and all the members of that one body, being many, are one body: so also is Christ." The Christian congregation and the army corps are made up of human bodies of particular types of disposition, brought together as a unity at particular times through cognitive and affective orientations. Indeed, the role of feeling and emotion in the creation of a corporate or collective body is through engaging aspects of the individual body in a particular and direct manner. Thus prayer and song in the Church, and route marches and drill in the army function to generate relevant feelings and emotions by and at the same time through engaging particular aspects of the body in a particular manner, depending on the organization in question.

Further, if the body is to be fully understood as a social phenomenon, then it is necessary to avoid a conceptualization of it which draws exclusive attention to the subject individual. Rather, it is important to conceptualize the body in a manner which directly refers to the interactive and relational, and therefore social (as opposed to socialized), aspects of the body. Instead of conceptualizing the bodies of individual persons in terms of the social constraints to which they are subjected, which focus attention on individual characteristics as individual propensities, it is more appropriate to emphasize the active bases of the embodied agent in relational and social forms.

The interactive dimension of agency acquires a broader basis when the social actor is understood as an embodied agent. Anthony Giddens (1984: 36) has commented that the "body is the 'locus' of the active self, but the self is obviously not just an extension of the physical characteristics of the organism that is its 'carrier.'" This type of objection disappears when the body is conceived not in its individual, biologically bounded form, but extended to include its social relations. For instance, Randall Collins (1981: 995) has said that for those who participate in it "social order must necessarily be physical and local," and also that the physical world includes "everyone's own body." By extension, social order pertains to the other bodies which are in some relation with one's own.

Following this type of reasoning it is possible to characterize particular social institutions in terms of the forms of bodily relations which they entail and which in turn make them possible. Thus Marx, for instance, accounts for capitalist factories in terms of the way workers utilize the "natural forces of [their] bod[ies]" in production, and find their bodies fragmented by the "detail dexterity" of factory work (Marx 1979 [1867]: 173, 340). This type of account allows us not only to characterize social institutions as particular types of relations between particular dispositions of bodies, but to see these bodily relations as the basis of the foundation and reproduction of these institutions. It is of interest that Collins (1981: 995) says, for example, that "[t]he most repetitive behaviors that make up the family structure are the facts that ... the same men and women sleep in the same beds and touch the same bodies, that the same children are kissed, spanked and fed." In this way it is possible to characterize different social institutions in terms of the distinct engagement of aspects of bodily disposition which pertain to them.

It can be seen, then, that bodies in social relationships, what can be called the "social body," are a complex of differentiated and simultaneous relationships between distinct aspects of individual bodies, depending on the social relationship or institution in question. The social body in this sense can be thought of as something like a somatized Dahrendorfian "homo sociologicus" (Dahrendorf 1968: 19–87). The body in society can therefore be seen as an ordered aggregation of particular and specific types of relationships between individual bodies or, rather, the relevant aspects of individual bodies.

The biological implications of social embodiment should not go unnoticed. For example, physiological changes brought about by co-presence and interaction have been observed and reported in the research literature of a number of disciplines including psychology, sociology, anthropology, and immunology (e.g. Kiritz & Moos 1974; Barchas 1976: 316–17; Ekman 1984; Freund 1988, 1990).

In the depiction of the body in society as "more than individual" and as institution-making, we move away from the body as merely subject to domination and passive. The body is intercommunicative and active. In a significant social sense the body referred to here is not simply the body we have but the body we are. The depiction of the social body above serves to overcome the tension between having and being a body, as treated by Helmuth Plessner (1970 [1941]) and taken up by Peter Berger and Thomas Luckmann (1967: 68), and in effect referred to by Giddens in the quotation above. Emotion is precisely the means whereby human bodies achieve a social ontology through which institutions are created.

The idea that the body is active in making its social world is given force and meaning through the idea that active bodies are also emotional bodies; that emotion is embodied. In general terms emotion is a concept which refers to the sense, including bodily sense, of evaluating experience. As Klaus Scherer (1984: 296) puts it, "one of the major functions of emotion consists in the constant evaluation of external and internal stimuli in terms of their relevance for the organism and the preparation of behavioral reactions which may be required as a response to those stimuli." The link between emotion and action is complex and certainly not mechanical. The role of emotion in various aspects of sociality such as associative behavior, group formation and integration, and the dynamics of social differentiation is treated in a growing literature (Collins 1984, 1990; Kemper 1984; Scheff 1988). Indeed, the idea that emotion is implicated in human social agency or praxis has its antecedents in the thought of Ludwig Feuerbach and Karl Marx who argued that it is through their senses and bodily activity that people feel, and it is through their emotions that people's activity has practical direction and force (Barbalet 1983: 31–62). These writers in particular encourage the view that the body is implicated in social agency through the facilitating role of emotions.

What we wish to emphasize here is that emotion is not only embodied but also essentially social in character. As Richard Lazarus, James Coyne and Susan Folkman (1984: 230) suggest, emotion is best regarded not as an "inner thing" but as a "relational process." Even at the organismic level, it cannot be conceived as a discrete psycho-physiological phenomenon. Both the limbic system and the neocortex are associated with the experience and expression of emotion. The function of each area can be seen to be integrated in the assessment of stimuli and choice of course of action (Pribram 1984: 30–2). These are precisely the areas associated with social behavior (Emde 1984; Pribram 1984).

The social relations implicated in emotional experience involve the body: not simply the body as a physical entity subject to external forces, but the body as agent. The distinction between having and being a body referred to above is in effect drawn by Levy (1984) in his discussion of Tahitian emotion. Without necessarily subscribing to the way in which he distinguishes between "feeling" and "emotion," it is interesting to observe that Levy shows that emotion engages the body not as a subject of self but as self; the self is embodied. In Levy's account feelings are an awareness which do not involve "the self" whereas emotions do (1984: 401–2). On this criterion, to say "my foot hurts" indicates a feeling (of pain), whereas to say "I am in pain" indicates an emotion. What is of interest here for our purposes is that the feeling-state described by Levy refers to the body (or part of the body – the foot) as a possessed "mine": I have a foot (or body), whereas the emotion statement refers not to possession but to self: I am my body.

It is important to be aware of the different implications of different conceptions of emotion. Those accounts which focus on cultural presuppositions, vocabularies of emotion, and other cultural constructions (e.g. Hochschild 1979; Shott 1979; Lutz 1988) suppose that social relationships are less important to emotional experience than the culturally given cognitive structures in terms of which social relations are supposedly given their meaning. In this approach the bodily basis of emotion is generally ignored or even denied. Accounts in which social relations are the basis of emotion (Barbalet 1992; de Rivera and Grinkas 1986), in which particular configurations of practical power or status, for example, are associated with particular patterns of emotional expression, offer a very different appreciation of the bodily dimension of emotion. In this latter perspective bodily processes are centrally implicated in emotion, as the social-relational formation of emotion is a relation of embodied agents (de Rivera 1977; Kemper 1978; Scheff 1988).

The concept of emotion required for the idea of embodied agency makes explicit what is too frequently left implicit in the scholarly discussion of emotion. This conception requires two qualities. One is the notion that emotion is socially efficacious, that is, that emotion has social consequences. The other is the idea that emotion has a social ontology, that emotion as an experience involving physical and phenomenal aspects has simultaneously a social-relational genesis. This is important because the emotional basis of the social body as it has been described above is in a significant sense founded in the structure or pattern of social relationships.

The role of emotion in social embodiment

The central role of emotion in embodied social life is an ontological issue, the foundations of which can be outlined by a consideration of sources in ethology and phenomenology. In *The History of Manners* (1978 [1939]) Elias wrote on the means by which social conditioning of the body occurs. Much later he returned to the question of emotion as an element in the "civilizing process" in an article on the origin and function of emotion in human society (1987). In this he opposes nature and culture and searches for the "hinge" connecting them. He locates this in the interrelationship between biological potential and the biological propensity for learning, a process which occurs in a social context (1987: 351). Drawing on the communicative functions of emotion, Elias makes the obvious and well known but none the less important point that while emotional capacities are "innate," emotional expression occurs in, and the linguistic and cultural categories of feeling are learned in, the context of human social life (ibid.: 361). Elias's argument, however, is primarily behaviorist, treating emotional capacities as modified by the social context, and he does little to explore the role of emotion in the social agency of the body.

The communicative and social functions of emotion are, of course, true of primates in general, and a consideration of the origins of affectivity, and its continuities between human and non-human primates is instructive. Emotion is not a central topic in ethology; it is primarily dealt with in terms of its role in motivational systems of behavior, such as in the way fear might be regarded as the motivational basis for a particular observed behavior. But within the ethological literature generally and in work on emotion influenced by ethology, the evolutionary foundations of emotion have been addressed. Such material provides a basis for the better understanding of the role of emotion in social embodiment (Emde 1984; Plutchik 1984; Pribram 1984; Reynolds 1981).

The idea, typically associated with nineteenth-century social evolutionary thought, that emotion is a remnant of more "primitive" capacities somehow progressively submerged by the development of reason, is still prevalent. It is an unacknowledged element in much social science and continues to exert influence, implicitly at least, in social theory and is responsible in part for its cognitive bias. The issue of the primacy of emotion in cognition and the role of cognition in emotion has excited a recent debate in the psychological literature on emotion (Lazarus 1984; Zajonc 1980, 1984).

The central place of the emotional expressive system in primate evolution and its importance and further elaboration in humans has been addressed by an ethologist, Peter Reynolds (1981). He sees any theory of human evolution which presupposes "the development of reason at the expense of emotion or of learning at the expense of instinct" as counter to the behavioral evidence (ibid.: 36). Human evolution, he states, "does not climb the 'ladder of reason'" (1981: 38).

Reynolds, following Pribram, functionally differentiates the cognitive mechanisms evolved for the assessment of internal and external consequences of action into two main types which he terms the affective and the instrumental modalities of action (ibid.: 80). These affective and instrumental modalities have evolved simultaneously, and, according to Reynolds, are functionally interlocked: "both can call upon learned and innate behavior and in man, at least, both are active simultaneously" (ibid.: 80). The two function together in a complex and integrated, non-hierarchical relationship: "This conception of the relationship between reason and emotion as one of integration rather than of subordination cannot be over stressed, for it constitutes one of the great liberating concepts of ethological science, a concept that stands in marked opposition to the major emphasis of Western thought" (Reynolds 1981: 81).

What is crucial here for our purposes is that affective systems have become "specialized for the mediation of social relationships" (ibid.: 82). Reynolds goes on to say that affect or emotion should not be regarded as expressive of private and idiosyncratic states, but should be understood "as the phylogenetically specialized channel by which internal states are made public" (ibid.: 82). From this perspective it becomes clear why human affective communication "does not lapse into disuse, as the Ladder of Reason would predict" (ibid.: 82). It is for these reasons that affectivity and instrumentality cannot be regarded as distinct or opposite but are necessarily functionally integrated in directing social behavior.

What is not developed in Reynolds but which must be made explicit is that emotion is necessarily embodied. A full although varied account of the body as agent in world experience, its "being-in-the-world," is in the phenomenological literature and especially in the work of Merleau-Ponty. This phenomenological perspective has been recently taken up by a number of anthropologists (e.g., Jackson 1981; Csordas 1990). However, our position extends this phenomenological notion of the body as an active agent in world construction through the consideration of the role of emotion in the process of social embodiment.

The phenomenological literature varies enormously in what role it assigns to the body in the generation of the "subjective matrix of experience." Merleau-Ponty was perhaps the most explicit in his acknowledgement of the importance of the body as agent of experience in his work on the phenomenology of perception (1962 [1946], 1964), although the importance of the body was not ignored in the work of other phenomenologists (e.g., Husserl, Marcel, Scheler, and Schilder).

Merleau-Ponty grounded perception in the experienced and experiencing body. The world as perceived through the body was, for him, the ground level for all knowledge, for it is through the body that persons gain access to that world. This bodily being is experienced through spatial and motor functions, in speech, language, gesture. This being present in the world also includes the incarnation in a body in both space and in time, i.e., in history and culture, and thus involves the ever-present experience of social being (Merleau-Ponty 1962 [1946]: 56–7; 360–5). Persons discover the social world, he states, "not as an object or sum of objects, but as a permanent field or dimension of existence (ibid.: 362). One "may well turn away from it, but not cease to be situated relatively to it. Our relationship to the social is, like our relationship to the [natural] world, deeper than any express perception or any judgement" (ibid.: 362).

Writing before Merleau-Ponty, Max Scheler could be seen to prefigure a partial development of the points made above in his work on the phenomenology of feeling (e.g., 1954 [1913]; 1961 [1912]). Scheler is mentioned here because the main aim of his phenomenology of feeling was the very breaking down of the Cartesian distinction between reason and emotion. In reacting against the rationalist condemnation of feeling as mere subjectivity, he attempted to show that there are feelings which have an "objective" character as opposed to the more personalistic idea of internal "subjective" feeling (Spiegelberg 1982: 293), and that these "objective" feelings occur in a social context (see, for example, Scheler 1954 [1913]: xlvii–xlviii, 8–50). In a sense, what Scheler was grappling with, in our terms, was "social emotion," the need to be able to represent the feeling dimension of "being-in-the-world."

Classic works on the senses and perception in phenomenological psychology provide fruitful examples for the further grounding of the agency of the body in social life. Such work, too, as in the case of the phenomenology of Merleau-Ponty, can be extended through a consideration of the role of emotion in this process. Perceptual psychology is concerned with what the perceiving organism brings to perception. James Gibson's work, for example, on the functions of the haptic system in the apprehension of the world (1962, 1966), can be seen to complement the use we have already made of material drawn from ethology and from phenomenology more generally.

The haptic system, from the Greek term meaning "able to lay hold of," is, according to Gibson (1966: 97), "[t]he sensibility of the individual to the world adjacent to his body by the use of his body." It is the apparatus through which information about both the body and the environment are gained (ibid.). The notion of the haptic system, including the concept of haptic touch, implies its active nature: "The haptic system, unlike the other perceptual systems, includes the whole body, most of its parts, and all of its surface. The extremities are exploratory sense organs, but they are also performatory motor organs; that is to say, the equipment for feeling is anatomically the same as the equipment for doing" (Gibson 1966: 99). Gibson's "feeling," of course, refers here to touch, rather than emotion. However, the haptic system must be seen as part of a dynamic and social process, for "in social touch, the haptic system with all of its subsystems comes into full perceptual use" (ibid.: 134). The work on haptic touch is useful in developing a sense of the agency of the body in both individual and social existence, and may thus contribute to the elaboration of the model of embodied feeling central to the argument of this chapter.

The role of feeling in social embodiment is more clearly acknowledged in the work of Heinz Werner and his associates. Werner and Wapner (1949) developed the "sensory-tonic field theory of perception" in order to depict perception as a "total dynamic process," in which the relationship of the organism to context is given full consideration, including "projective" characteristics of perception such as emotion (ibid.: 91). Later, in his work on symbolization, Werner writing with Kaplan (1978: 477) argues that the development of human symbolic capacities incorporates affective and sensorimotor response patterns of earlier (pre) symbolic behavior, but in ways which result in new cognitive, functional elaborations of these patterns (see also Werner and Kaplan 1963). Kaplan, in a later work, sees the primary form of symbolization, in which "abstract intangible states of affairs are realized in a concrete medium" (e.g., in mythopoetic thought), as associated with a sense of actuality and intensity of feeling (1979: 223).

A more explicit and particularistic concern with the embodiment of emotion is apparent in current research on the neurophysiological foundations of emotion (e.g., Ekman et al. 1983; Papanicolaou 1989) and on the integration of cognitive and emotional functions (e.g., LeDoux 1989). This work is important in the understanding of the interrelationships of affect, learning, and memory, but also the further understanding of practical action (for memory see, for example, Bower 1980; Casey, 1987; Bowles 1988).

Thus, the embodied agent cannot be treated simply in terms of how his or her experience of the world forms the basis of knowledge. This is not sufficient to explain action. Feeling must be integrated in an account of both the experience of the world and the understanding of action within it. An understanding of emotion and its foundation in sociality is part of and makes sense of embodied experience, and in turn locates within the body the basis for its agency in the world. Thus, emotion is essential to any conception of social life, as a link between embodiment on the one hand and the practical activity of social life, that is, the praxis of the body, on the other.

Conclusion

The interactive and relational aspects of emotion are, in fact, as etiologically crucial and constitutive as the psychological and cultural. The supra-individual and social-relational bases of emotion are unavoidable. The suggestion here is that the concept of embodiment need not reinforce the idea that the individual is necessarily to be understood in terms of his or her "internal" processes, but rather that the concept of embodiment itself can lead to a deeper awareness of the sociality of being and emotion. Further, the approach to embodiment developed here, in that it provides an understanding of social agency in terms of the sensory and affective foundations of embodiment and therefore of experience, indicates by extension the interdependency of the social with the biological foundations of the experiencing organism.

What has been emphasized in this chapter is how we may conceptualize the "social body," that is, how the body can be understood simultaneously at an individual and a social level, and as agent as well as object in the social world. We have discussed the crucial role of emotion in the being of the body in society. Our goal has been to develop a framework for seeing the body as agent in, and as locus of intersection of, both an individual psychological order and a social order, as well as for seeing the body as both a biological being and a conscious, experiencing, acting, interpreting entity. We have argued that it is through the agency of emotion that this synthesis is accomplished and may be understood, for it is through emotion that the intersection of individual order and social order may be most clearly seen.

REFERENCES

Barbalet, J. M. (1983) *Marx's Construction of Social Theory*. London: Routledge and Kegan Paul.
  (1992) A Macrosociology of Emotion: Class Resentment. *Sociological Theory* 10: 150–63.
Barchas, Patricia R. (1976) Physiological Sociology: Interface of Sociological and Biological Processes. *Annual Review of Sociology* 2: 299–333.
Berger, Peter L. and Thomas Luckmann. (1967) *The Social Construction of Reality: A Treatise on the Sociology of Knowledge*. London: Allen Lane.
Bower, Gordon C. (1980) Mood and Memory. *American Psychologist* 36: 129–48.
Bowles, Edmund Blair (1988) *Remembering and Forgetting: Inquiries into the Nature of Memory*. New York: Walker and Company.
Casey, Edward S. (1987) *Remembering: A Phenomenological Study*. Bloomington, IN: Indiana University Press.
Collins, Randall (1981) On the Microfoundations of Macrosociology. *American Journal of Sociology* 86: 984–1014.
  (1984) The Role of Emotion in Social Structure. In Klaus R. Scherer and Paul Ekman, eds., *Approaches to Emotion*. Hillsdale, NJ: Lawrence Erlbaum and Associates, pp. 385–96.
  (1990) Stratification, Emotional Energy, and the Transient Emotions. In Theodore D. Kemper, ed., *Research Agendas in the Sociology of Emotions*. Albany: State University of New York Press, pp. 27–57.
Csordas, Thomas J. (1990) Embodiment as a Paradigm for Anthropology. *Ethos* 18(1): 5–47.
Dahrendorf, Ralf (1968) *Essays in the Theory of Society*. Stanford: Stanford University Press.
de Rivera, Joseph (1977) *A Structural Theory of the Emotions, Psychological Issues Monograph 40*. New York: International Universities Press.
de Rivera, Joseph and Carmen Grinkas (1986) Emotion as Social Relationships. *Motivation and Emotion* 10: 351–69.
Durkheim, Emile (1954) [1912] *The Elementary Forms of the Religious Life*. Glencoe, IL: Free Press.

Ekman, Paul (1984) Expression and the Nature of Emotion. In Klaus Scherer and Paul Ekman, eds., *Approaches to Emotion*. Hillsdale, NJ: Lawrence Erlbaum and Associates, pp. 319–43.

Ekman, P., R. W. Levenson, and W. V. Frieson (1983) Autonomic Nervous System Activity Distinguishes Between Emotions. *Science* 221: 1208–10.

Elias, Norbert (1978) [1939] *The Civilizing Process: Sociogenetic and Psychogenetic Investigations. Volume I: The History of Manners*. Oxford: Basil Blackwell.

(1987) On Human Beings and Their Emotions: A Process-Sociological Essay. *Theory, Culture and Society* 4: 339–61.

Emde, Robert N. (1984) Levels of Meaning in Infant Emotions: a Biosocial View. In Klaus R. Scherer and Paul Ekman, eds., *Approaches to Emotion*. Hillsdale, NJ: Lawrence Erlbaum and Associates, pp. 77–107.

Featherstone, Mike (1982) The Body in Consumer Culture. *Theory, Culture and Society* 1: 18–33.

Foucault, Michel (1979) *Discipline and Punish*. Harmondsworth: Penguin.

(1980) *The History of Sexuality. Volume 1: An Introduction*. Harmondsworth: Penguin.

Foss, Lawrence and Kenneth Rothenberg (1988) *The Second Medical Revolution: From Biomedicine to Infomedicine*. Boston: New Science Library.

Frank, Arthur W. (1990) Bringing Bodies Back In: A Decade Review. *Theory, Culture and Society* 7: 131–62.

Freund, Peter E. S. (1988) Bringing Society into the Body. *Theory and Society* 17: 839–64.

(1990) The Expressive Body: A Common Ground for the Sociology of Emotions and Health and Illness. *Sociology of Health and Illness* 12: 452–7.

Gibson, James J. (1962) Observations on Active Touch. *Psychological Review* 69: 477–91.

(1966) *The Senses Considered at Perceptual Systems*. Boston: Houghton Mifflin Co.

Giddens, Anthony (1984) *The Constitution of Society: Outline of a Theory of Structuration*. Cambridge: Polity Press.

Haraway, Donna J. (1991) *Simians, Cyborgs and Women: The Reinvention of Nature*. New York: Routledge.

Honneth, Axel and Hans Joas (1988) *Social Action and Human Nature*. Cambridge: Cambridge University Press.

Hochschild, Arlie Russell (1979) Emotion Work, Feeling Rules, and Social Structure. *American Journal of Sociology* 85: 551–75.

Jackson, Michael (1981) Knowledge of the Body. *Man* 18: 327–45.

Kaplan, Bernard (1979) Symbolism: From the Body to the Soul. In Nancy R. Smith and M. B. Franklin, eds., *Symbolic Functioning in Childhood*. Hillsdale, NJ: Lawrence Erlbaum and Associates, pp. 219–28.

Kemper, Theodore (1978) *A Social Interaction Theory of Emotions*. New York: John Wiley and Sons.

(1984) Power, Status and Emotions: A Sociological Contribution to a Psychophysiological Domain. In Klaus R. Scherer and Paul Ekman, eds., *Approaches to Emotion*. Hillsdale, NJ: Lawrence Erlbaum and Associates, pp. 369–83.

Kiritz, S. and R. H. Moos (1974) Physiological Effects of Social Environment. *Psychosomatic Medicine* 36: 96–114.

Lazarus, Richard S. (1984) Thoughts on the Relations Between Emotion and Cognition. In Klaus R. Scherer and Paul Ekman, eds., *Approaches to Emotion.* Hillsdale, NJ: Lawrence Erlbaum and Associates, pp. 247–57.

Lazarus, Richard S., James C. Coyne and Susan Folkman (1984) Cognition, Emotion and Motivation: The Doctoring of Humpty Dumpty. In Klaus R. Scherer and Paul Ekman, eds., *Approaches to Emotion.* Hillsdale, NJ: Lawrence Erlbaum and Associates, pp. 221–37.

LeDoux, Joseph E. (1989) Cognitive-Emotional Interactions in the Brain. *Cognition and Emotion* 3: 267–89.

Levy, Robert I. (1984) The Emotions in Comparative Perspective. In Klaus R. Scherer and Paul Ekman, eds., *Approaches to Emotion.* Hillsdale, NJ: Lawrence Erlbaum and Associates, pp. 397–412.

Lutz, Catherine (1988) *Unnatural Emotions: Everyday Sentiments on a Micronesian Atoll and Their Challenge to Western Theory.* Chicago: University of Chicago Press.

Lyon, Margot L. (1990) Order and Healing: The Concept of Order and Its Importance in the Conceptualization of Healing. *Medical Anthropology* 12(3): 249–68.

(1993) Psychoneuroimmunology: The Problem of the Situatedness of Illness and the Conceptualization of Healing. *Culture, Medicine and Psychiatry* 17(1): 77–97.

Marx, Karl (1979) [1867] *Capital. A Critical Analysis of Capitalist Production. Volume 1.* Moscow: Progress Publishers.

Merleau-Ponty, Maurice (1962) [1946] *The Phenomenology of Perception,* Colin Smith, trans. London: Routledge and Kegan Paul.

(1964) *The Primacy of Perception,* J. Edie, ed. Evanston, IL: Northwestern University Press.

O'Neill, John (1985) *Five Bodies: The Human Shape of Modern Society.* Ithaca: Cornell University Press.

Papanicolaou, A. C. (1989) *Emotion: A Reconsideration of the Somatic Theory.* New York: Gordon and Breach.

Plessner, Helmuth (1970) [1941] *Laughing and Crying. A Study of the Limits of Human Behavior.* Evanston, IL: Northwestern University Press.

Plutchik, Robert (1984) Emotions: A General Psychoevolutionary Theory. In Klaus Scherer and Paul Ekman, eds., *Approaches to Emotion.* Hillsdale, NJ: Lawrence Erlbaum and Associates, pp. 197–219.

Pribram, Karl H. (1984) Emotion: A Neurobehavioral Analysis. In Klaus Scherer and Paul Ekman, eds., *Approaches to Emotion.* Hillsdale, NJ: Lawrence Erlbaum and Associates, pp. 13–38.

Reynolds, Peter C. (1981) *On the Evolution of Human Behavior: The Argument from Animals to Man.* Berkeley: University of California.

Scheff, Thomas J. (1988) Shame and Conformity: The Deference-Emotion System. *American Sociological Review* 53: 395–406.

Scheler, Max (1961) [1912] *Ressentiment.* New York: The Free Press.

(1954) [1913] *The Nature of Sympathy.* London: Routledge and Kegan Paul.

Scheper-Hughes, Nancy and Margaret M. Lock (1987) The Mindful Body: A Prolegomenon to Future Work in Medical Anthropology. *Medical Anthropology Quarterly* n.s. 1: 6–41.

Scherer, Klaus (1984) On the nature and function of emotion. In Klaus R. Scherer and Paul Ekman, eds., *Approaches to Emotion.* Hillsdale, NJ: Lawrence Erlbaum and Associates, pp. 293–317.

Shott, Susan (1979) Emotions and Social Life: A Symbolic Interactionist Analysis. *American Journal of Sociology* 84: 1317–34.

Spiegelberg, Herbert (1982) *The Phenomenological Movement: A Historical Introduction.* The Hague: Martinus Nijhoff.

Turner, Bryan S. (1984) *The Body and Society: Explorations in Social Theory.* Oxford: Basil Blackwell.

Werner, Heinz and Bernard Kaplan (1963) *Symbol Formation: An Organismic–Developmental Approach to Language and the Expression of Thought.* New York: John Wiley and Sons.

(1978) The Nature of Symbol Formation and Its Role in Cognition: Theoretical Considerations. In S. S. Barten, M. B. Franklin, eds., *Developmental Processes: Heinz Werner's Selected Writings.* New York: International University Press, pp. 471–85.

Werner, Heinz and Seymour Wapner (1949) Sensory-Tonic Field Theory of Perception. *Journal of Personality* 18: 88–107.

Wrong, Dennis (1961) The Oversocialized Conception of Man in Modern Sociology. *American Sociological Review* 26: 183–93.

Zajonc, R. B. (1980) Feeling and Thinking: Preferences Need No Inferences. *American Psychologist* 35: 151–75.

(1984) On Primacy of Affect. In Klaus Scherer and Paul Ekman, eds., *Approaches To Emotion.* Hillsdale, NJ: Lawrence Erlbaum and Associates, pp. 259–70.

Part II

Form, appearance, and movement

3 The political economy of injury and compassion: amputees on the Thai–Cambodia border

Lindsay French

This chapter is about the influence of power relations on the experience and understanding of bodily harm. More specifically, it is about amputees on the Thai–Cambodia border, and the effects that almost two decades of hunger, terror, war, and long-term confinement have had on a population's experience of their own and other people's damaged bodies. It begins with a model for considering the phenomenal world in terms of a series of domains in which the experiencing self is involved, then extends this model to include the domain of power and its social relations. My subject is a population of Cambodians who have survived the depredations of Pol Pot's rule and sustained an unusually high number of landmine injuries over the course of a twelve-year guerrilla war that followed the overthrow of the Khmer Rouge regime in 1979.[1] My thesis is that even the most apparently subjective and personal experiences – the experience of one's own body – is shaped in important ways by the relations of power and domination in which the body is involved. These relationships are embedded in the social order and are part of the experience of everyone who participates in that order. As such, they contribute to the shaping of the lived world of the entire population.

In October 1991, an agreement was signed in Paris that brought to a close the most recent chapter in an ongoing history of violent conflict in Cambodia, a history of violence which had scarcely ceased in twenty-two years. The peace accords officially ended a guerrilla war that had been waged since 1979 against the Vietnamese-backed government in Phnom Penh, by three different resistance armies operating out of camps on the Thai–Cambodia border. This agreement was a hard-won and important step on the tortuously twisted path toward peace in Cambodia. The most insidious weapon of the war, however, and the weapon that has arguably caused the most social and psychological damage to the population, remains intact and will continue to claim victims for many years to come. That weapon is the anti-personnel landmine.

Landmines have been used extensively by all parties in this guerrilla conflict. They are small, inexpensive, easy to set, difficult to detect, and they remain active in the forests and rice paddies of Cambodia for decades. They

3.1   Street scene. Camp for displaced Khmer on Thai–Cambodia border. (© Bill Burke, Swanstock.)

are strategically effective because they disengage several soldiers from active combat to attend to each soldier wounded by one, and they terrify the civilian population by exploding in fields and forests long after the soldiers who set them have left the area. They are the signal weapon of this war, and their signature has been indelibly inscribed on the bodies of the Cambodian people. Widely spread and often deliberately set with just enough explosives to maim but not kill their victims, they have left their mark in amputations among civilians and soldiers alike.

In the population of Cambodians with whom I worked between 1989 and 1991, a camp for civilians displaced by the war which was associated with one of the guerrilla factions,[2] the effects of the landmines were plainly visible. It was scarcely possible to find a street scene or neighborhood gathering that did not include at least one amputee: a soldier or civilian who had survived a landmine explosion but had lost an arm or a leg in the process. Amputees were ubiquitous in Site II: on the streets, in the markets, or sitting in front of their houses on the bamboo platforms that serve triple-duty as beds, tables and chairs. They had become a part of the landscape at the border, the lived world of this displaced population.

Amputation alters the integrity of the body in a particularly powerful way that affects not only the amputees themselves but also, in a different way, anyone who comes into contact with them. Our own intimate involvement with our bodies and our bodies' involvement in everything we do generates a complicated response to such dramatic disfigurement of another person's body. It is a striking and unusual situation to have so many mutilated bodies in one population and something, I thought, which must have profound consequences for the experience of all the people involved. This chapter began as an enquiry into that experience: what effect *did* all these amputations and amputees have on the population in Site II? How were these damaged bodies experienced? How were they "read"? What did it do to a group of people to have so many bodies like this in their midst?

My initial orientation to this question was "psychological" and "cultural." I was thinking in terms of "experiencing bodies": of the cultural construction of self and the different ways that affective, intellectual and physical awareness can be integrated, together with the broader question of how a whole population is affected by so much bodily trauma and disfigurement. What I found was a set of responses that could not be understood without reference to the political, economic and historical context of the amputees, however, as well as to a Theravada Buddhist hierarchy of virtue in which the significance of karma and of whole, intact bodies were centrally relevant. The "experience" could not be separated from this. What follows is an effort to show how the experiential world of the amputees is shaped by relationships of power and domination in Site II. My analysis focuses on the political economy of damaged bodies in one camp at a particular moment in Cambodian history, and on the ways in which this configuration of power relations constrains and encourages certain modes of affect and emotion. I hope to show that the specific cultural meaning of these bodies is not distinct from but deeply embedded in the relations of domination and production that have framed the war for these people.

The problem

Unlike the high-tech, postmodern wars of "smart" bombs, "Star Wars" missile battles, and "the attrition of critical levels of enemy material" (see Gusterson 1991), the war in Cambodia has been a small-scale, low tech-nology war that focused specifically on the mutilation, killing and display of bodies. Human bodies were at the center of all military action in this war, which was carried out in the jungles and rice paddies of Cambodia with AK-47's, M-16s, rocket grenade launchers, artillery, a few tanks, and, significantly, landmines. Although landmines were originally designed to be used defensively, to limit the military operations of the enemy, all parties in the Cambodian conflict have made extensive use of landmines as offensive weapons as well. The widespread and indiscriminate use of landmines is credited with the deaths of tens of thousands of soldiers and civilians since the war began in 1979 (Asia Watch 1991: 21; Handicap International 1990).

This casualty figure is matched by an even greater number of injuries. A 1991 Asia Watch report on the use of landmines in Cambodia estimates that there are now more than 30,000 amputees in a population of something over 8 million Cambodians, and 5–6,000 amputees living in camps on the Thai–Khmer[3] border (Asia Watch 1991: 1). Out of a total border population of approximately 350,000, this means that by conservative estimate one in every seventy people on the border in 1991 was an amputee.

These figures do not have the same kind of impact as seeing a man whose eyes were blown out by a landmine and now never speaks or leaves his hut, a woman who has lost both hands in an explosion that killed her husband, or a legless torso lying face down on a bamboo bed. These were among the more unsettling scenes of amputees I encountered in the camp where I worked. It was more common to see young men moving about the neighborhoods in bamboo wheelchairs or on handmade wooden and leather prostheses, or on crutches with no prosthesis at all.

There is something riveting about these images of amputees, perhaps because we all inhabit bodies and live so fundamentally *through* them, that brings the consciousness of bodily harm home to us. We respond viscerally to the specter of amputation: it challenges our own sense of bodily integrity, and conjures up nightmares of our own dismemberment. We feel an instinctive

the experience of dismemberment

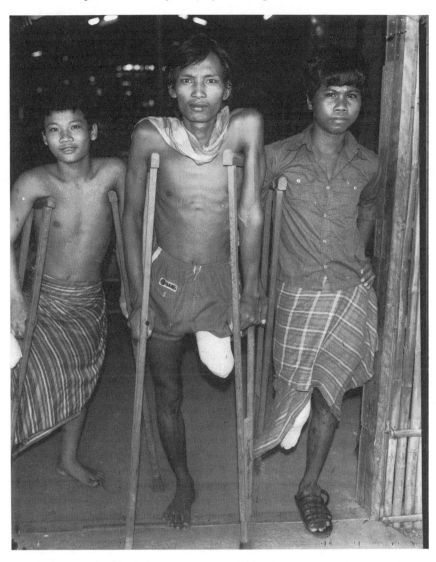

3.2   Recent amputees standing outside the surgical hospital that serves the entire Thai–Cambodia border. (© Bill Burke, Swanstock.)

sympathetic identification with the amputee by virtue of our own embodied being, but our identification frightens us; thus we are drawn toward and repelled by amputees simultaneously, both feeling and afraid to feel that we are (or could be) "just like them."

This generalized capacity for sympathetic identification with other "bodies" (however conflicted in the case of amputated bodies) is part of what makes the human body such a powerful instrument for communication. Indeed, images of amputees were used in the publicity material of virtually all organizations working with victims of this war, including those of the Khmer themselves. Allen Feldman, in another context (1991: 8), writes of "the mobilization of values through the spectacle of the body" wherein "the individual body is constructed as a mass article and as a social hieroglyph that opens the possibility of mythic communication with the masses." Feldman is drawing attention to the semiotic aspect of the body here, and its powerful communicative potential. His description is very apt, for it is on precisely this "mythic" level that the "spectacle" of the amputee communicates with us: we feel a decontextualized (visceral) kind of identification with the body of another person.[4]

But the body is also *experienced*, it is attached to a person in a social context, it is a subject as well as a semiotic object. Perhaps because we *feel* our embodied but decontextualized identification, we are liable to confuse this with understanding the experience of the other. But in fact, though we may believe we understand something through our instinctive, visceral identification with the body of an amputee, there is a contextual specificity to the meaning of bodily harm that makes it very difficult to understand the experiential dimensions of another person's injury.[5] This became very clear when I discovered how wrong my initial assumptions about the meaning of amputation in Site II really were.

When I began interviewing for this chapter I expected to learn something new about living close to war, and about the anxiety I knew went along with this experience. I imagined, incorrectly, that the amputees would constitute a kind of inexorable reminder of the war, but that their physical presence would also generate some sort of compassionate Buddhist response. What I found was strikingly different. Amputees in Site II elicited not a generalized anxiety about war but a quite specific anxiety about personal safety: people were afraid of them. Amputees, in particular young male amputees, had acquired a reputation for violence, extortion, and theft, and tended to be given a wide berth. Amputees were almost universally looked down upon, but rarely regarded with compassion; *their* central experience was one of degradation and abandonment, by their leaders, their families, and the society as a whole.

How was this situation to be understood?

The more I worked with these responses the more the data suggested beginning with an analysis of the political economy of injured bodies such as Foucault developed in detail in *Discipline and Punish*. Foucault reminds us that whatever physiological or sensuous qualities the body may hold for us, it is also "involved in the political field"; that "the political investment of the body is bound up in accordance with complex reciprocal relations, with its economic use"; that "it is largely as a force of production that the body is invested with relations of power and domination, but on the other hand, its constitution as labor power is caught up in a system of subjection (in which it is also a political instrument meticulously prepared, calculated, and used); the body becomes a useful force only if it is both a productive body and a subjected body" (1979: 25–6).

My analysis of the meaning of injured bodies in Site II turns on the ways in which bodies were both productive and subjected in this political, historical, and cultural context, and how this situation was experienced by the people who lived there. My project is phenomenological in that it focuses on lived experience and its meanings, but it locates these meanings in the intersubjective (and often contested) space of social relations and cultural signs – what the anthropologist A. I. Hallowell calls the "behavioral environment." At issue are subjectivity (the meaning of lived experience), the "subject" (amputees, and the people with whom they live), and how these are best to be understood. I will argue that subjective experience is inadequately conceived if it does not take into account the subject's involvement in relations of production and power (its *subjection*); that subjectivity is inseparable from the intersubjective and material dynamic of these relations.

Theoretical considerations

Phenomenologists since Husserl have been concerned with understanding the relationship between the lifeworld of phenomena that are "given" to us to experience and our apprehension and understanding of that lifeworld. Although different philosophers have taken Husserl's work in different directions, there has been an abiding interest in the investigation of the particular phenomena of the lifeworld, the general "essences" of phenomena, and the essential relationship among essences, as well as the philosophical basis of the "natural attitude" of being-in-the-world. (See Spiegelberg, 1984, and Natanson 1973, especially pp. 3–44.) Through these investigations has come an appreciation for the *social* nature of the lifeworld as it is constituted through our actions and intentionality. That is, while we experience as individuals, *what* we experience is constituted through our complexly interwoven subjectivities, and through our actions and interactions which are based on these intersubjectivities.

A. I. Hallowell follows in this philosophical tradition in his anthropological essay entitled "The Self and Its Behavioral Environment," published in 1955. Hallowell acknowledges his debt to phenomenology as he develops his own down-to-earth model for thinking about the relationship between the world outside the self and the subjective, experiencing self. In his essay Hallowell suggests that the world as experienced by any individual is shaped by several fundamental "orientations" of the self: to itself, to other objects outside itself, to the spacio-temporal dimension, to its motivational universe, and to the normative world (Hallowell 1955: 75–110). Through the combined influence of these different "orientations," the reflexive, experiencing self incorporates the external world into its sphere of understanding and action. The "objective" world is transformed into a "behavioral environment" within which certain actions are possible while others are ludicrous or dangerous, and certain emotions are meaningful while others make no sense. Through these basic orientations the "objective" world is endowed with significance; within this "behavioral environment" individual selves think, feel and act.

Hallowell's work focuses on the dynamic process of self-awareness in the lived world, or behavioral environment, and on the ways in which meaning is produced and reproduced socially within this world. His model pays specific attention to the shared aspects of the experienced world which derive from social and cultural processes, and is thus broad enough to accommodate the domain of power relations into the experiential realm. His own analyses, however, do not highlight this aspect of experience.

Foucault on the other hand has explored the dynamics of power with respect to the *body* in great detail. The "genealogy" he develops in his later works (see especially *The History of Sexuality, Volume I*, and *Discipline and Punish*) provides us with an analytic technique for understanding historical and cultural formations of power in relation to bodily experience and its meaning. Foucault's "analytics of power" are concerned with uncovering "the multiplicity of force relations immanent in the sphere in which they operate," and the body in both its objective and subjective aspects provides a "grid of intelligibility of this social order" (1980a: 93). His analytics demonstrate how we may think about social relations and processes through their orientation to power, and analyze these through their systematic effects on historical bodies.

Thus while Hallowell is concerned with the meaningful incorporation of the external world into the experiential realm of the self (and with the definition of this realm as an interactive behavioral environment), Foucault is interested in a more abstract level of coherence, in which even subjective bodily experience functions as a sign of complex power relations rather than constituting a locus of meaning in itself. "His interest," as Dreyfus and

Rabinow (1982: xxvii) have put it, "is in the social effects rather than the implicit meaning of everyday practices." For Foucault the point of carefully analyzing historical experiencing bodies is found in what they can reveal to us about these larger configurations of power, not in what the experience itself has to say. In this way his concerns differ radically from many phenomenologists' concerns with "essences."

My own approach attempts to bring together the analytic potential of Foucault's bodies with the phenomenological concerns of Hallowell's selves. My interest in the experiential world of Cambodian amputees focuses on the orientation of bodies/selves to power in the particularly constructed "behavioral environment" of Site II, and on the ways in which relationships of power and exploitation interact with material conditions and deep-seated religious orientations to shape a context for experience, understanding and action. I will try to combine Foucault's "emphasis on the body as the place in which the most minute and local social practices are linked up with the large scale organization of power" (Dreyfus and Rabinow 1982: xxii) with Hallowell's sensitivity to the interactive dimensions of the experiential realm, to get at the effects of political, economic, and historical factors on the experience of living in and among amputated bodies in Site II. I am less concerned here with providing a phenomenological account of individual experiences of amputation than I am with showing how relations of domination and production shape the intersubjective, social experience and understanding of amputation in the population as a whole.[6]

In Site II most of the amputees were former soldiers. Although land-mines are strewn throughout the war zone in such a way that many civilians inside Cambodia have been injured in the course of their daily activities, in the population with whom I was working civilians were confined to the camp, which was located just across the Cambodian border in Thailand. Some camp residents left the camp in spite of these regulations, and encountered landmines in the nearby forests where they went to collect vegetables or firewood. Others stepped on mines while travelling back and forth between the camp and Cambodia. But most of the victims were young soldiers, injured while on active military duty "inside." My analysis there-fore focuses on the injury to young male bodies.

The political economy of soldiering in this guerrilla war

Let me first contextualize these soldiers in the political economy of the war. This war began in 1979, when the Vietnamese army overthrew the murd-erous Khmer Rouge regime in Cambodia and created a new communist government out of renegade cadres from the original Khmer Rouge organi-zation and old Cambodian communists who had been living in Vietnam since the 1950s. A resistance was formed on the Thai–Cambodia border around the goal of ousting the Vietnamese from Cambodia; it consisted of supporters of the prince and former president Norodom Sihanouk, the remains of the routed Khmer Rouge, and a non-communist, non-royalist faction under the former Prime Minister Son Sann. These three factions were loosely linked in a strategic military coalition; organizationally and ideologically, they remained separate and distinct. The third faction, called the Khmer Peoples' National Liberation Front, or KPNLF, controlled the population in Site II.[7]

Many of the leaders of the KPNLF resistance had been active in political affairs in the years before the Khmer Rouge seized power in 1975, and had a genuine political commitment to the goals of the resistance. Most of the soldiers, who sustained the bulk of the injuries of war, had little such commitment, however. Most had come of age under the Khmer Rouge, had little or no formal education, and in the political uncertainty, displacement and war that followed the fall of the Khmer Rouge, had no reliable way of supporting themselves or their families. For many young men the decision to become a soldier was much more pragmatic than it was political. The army provided a kind of security in this profoundly unsettled time: they were given a role in an authority structure as well as a job. Their commitment was to a particular commander, who served as their patron: in exchange for loyal service, their commander provided uniforms and weapons, found a place for their families in a civilian refugee camp, and served as a resource to whom the soldiers could turn in times of trouble or need. The soldiers provided their commanders with a base of power, in exchange for which the commanders looked after their interests and gave them a political affiliation.

This unequal but reciprocal sense of obligation was an important aspect of the relationship as very few material resources actually changed hands. Foot soldiers were rarely paid for their services, nor were they even provided with adequate rations when they were out on military operations.[8] But under the conditions of scarce resources and uncertain authority that have characterized post-Pol Pot Cambodia, it was important for poor men to secure the patronage of someone more powerful than themselves. These soldiers spoke with a kind of unswerving loyalty about their commanders; on the other hand many said that the work was too hard and they would quit the army if they had anyone else who could "take care of" them. Most felt they had little practical choice but to offer themselves to the authority of these commanders.

In exchange for protection and patronage, soldiers provided their commanders with productive labor in the business of war. They trained, fought, occupied territory, took prisoners, liberated weapons from the enemy, etc.

But when they lost an arm or a leg, they were no longer a productive part of this economy. Their commanders had no use for them, and the resistance did not have the resources to support soldiers wounded in combat; that was the hard reality from the commanders' point of view. From the soldiers' perspective, however, their commanders had backed out of a moral obligation, for it was the need for support in precisely this kind of situation that got them into soldiering in the first place. One amputee soldier described the military patronage system in action this way: "Our commanders just need people who can do their work for them. When we cannot work anymore, they just throw us away."

These amputees were very bitter about their abandonment by the resistance, a response that comes at least in part from the nature of their initial commitment to their commanders. The commanders were willing to maintain a patronage relationship as long as they were productive soldiers. But in the economy of this war an amputee was worthless, and a soldier could do little to invoke his commander's support once he had lost a limb. What was undertaken by the soldiers as a moral relationship, that created and maintained reciprocal social bonds, had been transformed into something purely material and utilitarian, to be abandoned by their commanders when it lost its usefulness for them.[9]

The domestic economy

In the domestic economy of everyday life, an amputee's value was greatly diminished as well. Although Site II occupied a marginal and resource-poor plot of scrubland on the Thai–Khmer border, a complex social and cultural life had nevertheless developed in the seven years since the camp had been created, sustained primarily by United Nations assistance, black-market trade, and the economy of the war. With a population of approximately 180,000, Site II was the second largest city of Cambodians, after Phnom Penh. The UN provided subsistence rations and basic materials for building bamboo and thatch huts; beyond this people exchanged whatever goods and services they could generate in a cash-poor legal market inside the camp, and traded illegally on the cross-border black market outside. Outside the camp there was money to be made but it was dangerous for camp residents; unless they were under the authority of a KPNLF commander they were illegal in Thailand and subject to arrest. Inside the camp it was safer but much more difficult to make any money at all. It was within this restricted economic context that men and women endeavored to fulfill the largely traditional gender roles of "provider" and keeper of the house and household purse.

In the sexual division of labor Cambodian men are expected to provide the means for feeding and supporting their families and protect them from physical harm. For a single family living in a camp on the Thai–Khmer border, however, security was tenuous indeed and the threat of victimization by either the Thai "Protection Unit" or some other Cambodian was very real. The protection provided in the past by extended kinship networks and close-knit village life had been destroyed by years of war and repeated displacement. In Site II, mistrustful of neighbors whose history they did not know or dare to find out, families turned inward for protection. Often all that stood between them and robbery, extortion, kidnapping and even rape was an able-bodied adult male in the house. Often a family's reaction to the news that a husband or son had lost a leg was a kind of panic at the prospect of their new vulnerability: who would be able to protect and support them now? This raw fear suggests just how close to the edge of danger many people felt their lives in Site II to be.

In Site II an amputee's work options *were* limited in comparison to those of an able-bodied man. One amputee said to me, "When I had two legs I had the ability to work, no matter where I was. But when I lost my leg nobody needed me to do anything." Jobs were scarce in Site II, and many foot soldiers were ill-prepared for any work other than military service; indeed, it was often the lack of any employment alternative that propelled them into the army in the first place. Amputation often robbed these men of the only work they could do in this situation that had any status or "security" at all.

This was a hard blow for a man whose social worth is largely evaluated in terms of his ability to protect and provide for his family. Considerable shame accrues to a man who cannot provide his family with the resources to live at the level to which they are accustomed. Another amputee said to me,

My wife often complains when I cannot get money or food for her. Some handicapped men's wives abandon their husbands or go away with new husbands. And some wives go back to their parents in Cambodia. If we do not have money our children do not respect us, because they do not have money for their teacher's fees. Even my wife does not respect me so much now. If I cannot support her, maybe *she* will leave me too.

Thus shame, anger, and a sense of impotence as well as loss were combined for this amputee with the very real fear of losing his place in the most basic institution of social support for Cambodians, his own family. The interdependence of the domestic economy and the economy of war and their joint effect on this man's struggle to maintain a credible social identity are clearly demonstrated in his remarks.

The ideology of amputation

But there is something more going on here than just physical injury and the social disability this entails, or even the loss of the body as an economic resource, some more fundamental devaluation that has to do with the meaning of a body that is not complete or whole. Amputees often seemed overcome by an overall sense of incapacity and worthlessness; in some fundamental way they had been devalued by the loss of a limb. One man described his feelings this way:

Nowadays I feel strange, and different than when I was young. Right now my mind feels very weak; I am not clever like I used to be. Before I lost my leg I was a good student. No one was ever above me; I was always the first in my class. Now I am not so clever as I was before I lost my leg.

The Khmer term used most frequently in Site II to describe this complex loss of competence or ability was *samattaphiep*, which translates as "capacity" or "competence" but also includes the idea of "courage" (Headley 1977). Often this term was used to refer to a condition that involved a physical incapacity but had, as in the example cited above, much deeper implications for the injured party. That is, serious physical disability often carried with it a sense of diminished mental capacity, a loss of vitality or effectiveness, an overall personal weakness. The fact that Khmer understand health to be a complex, multi-faceted condition which involves emotional and spiritual as well as physical well-being provides some context for the use of this term. Conversely, physical terms were often used to express the diffuse, generalized feelings of malaise that were common in Site II. Khmer generally thought about any ailment in relation to an overall, integrated sense of health.[10]

But the amputee's overall feeling of diminished "capacity" has particular significance with respect to a Theravada Buddhist hierarchy of merit or virtue, and in relation to the concepts of karma and reincarnation. In the Buddhist universe there is a ranked continuum of sentient beings that stretches from the lowliest earthworm up through the animal kingdom to the kings and angels and bodhisattvas, and ultimately to the Buddha himself. Each being inhabits a different position in this hierarchy of virtue according to its karma, or destiny, which is the result of all of its actions, good and bad, stretching back through the history of all of its past lives. We accumulate merit through our moral actions, and suffer according to our karma. If our karma is good we will be born rich and powerful, but if it is bad we will be born poor or crippled or orphaned. These are both the signs and the manifestations of our karmic status.

Just to be born human is good fortune, but there are obvious hierarchies of status and virtue *among* humans, and an incomplete or crippled human body is benighted indeed. To be born able-bodied and healthy and then suddenly to lose a limb thus constitutes a fall in status, a drop in value. Perhaps as important, it represents a rapid downturn in one's fortune, a sudden and inauspicious ripening of one's karma, or destiny. It does not bode well for the future.

This sense of foreboding that many amputees felt runs somewhat counter to classic Theravada doctrine. According to the Buddha's teaching, the future is never fully determined. The future remains open regardless of one's karmic accumulations from the past because one can always influence one's destiny through meritorious acts in the present (Keyes 1983: 1–24). Moreover, while karmic misfortune "is seen as the consequence of moral actions in a previous existence," in most Theravada contexts there are "few if any connotations of personal responsibility ... rather ... karma is construed as an impersonal force – the law of karma – over which one has no control" (ibid.: 15).[11] Theravada doctrine counsels acceptance of the past which cannot be changed, and promotes a forward-looking, open attitude toward the future through mindfulness and right action in the present.

Traditionally Cambodians have invoked karma to account for circumstances and events that were beyond their ability to control: drought, flood, crop failure, an unforeseeable business loss, or the accidental death of a family member, for example. Cambodians have a fundamentally pragmatic outlook, and have traditionally accepted as karma those difficult things in their lives which they could do little to change. What was taken to be outside one's ability to control depended on one's social position, of course: a well-educated and wealthy businessman in Phnom Penh had a much greater sense of personal efficacy than a poor rice farmer from the provinces. But this in itself was an example of karma: wealthy people had good karma and were powerful; poor people didn't, and were not. For most people karma was a sufficient explanation for this, and life went on, bringing the joys and sorrows appropriate to one's place in the social hierarchy.[12] Meanwhile people put their intentionality to work in those areas of their lives in which they could be effective, sought supernatural protection from spirits and magic through the mediation of traditional *krus* [gurus, sorcerers, healers], and made merit against the future in the context of a local Theravada tradition, in which the annual cycle of Buddhist ceremonies helped to maintain the sense of an integrated universe of power and meaning.

But in Site II, that sense had been badly shaken. The crushing weight of two decades of suffering caused many people to lose faith in the possibility of a better future, and to experience their karma as irrevocable rather than open and subject to change. In spite of the Buddha's teaching many Cambodians did ponder why such bad fortune had been visited upon them, and on the Khmer people as a group. The sense of accumulating bad fortune was hard to reverse in the context of life in Site II, where in practical terms people had very little ability to influence their future in significant ways. They were, in essence, prisoners in the camp until a settlement was reached in the guerrilla struggle, and no settlement was in sight during the time I was recording people's responses. For many, each new misfortune was experienced as further confirmation of a hopelessly degraded karmic destiny. Thus amputation was often "read" as a "sign" of bad fortune.[13] In purely practical terms it foreboded a much more difficult life in the present for the amputee and his family as well.

Amputees in the social world

The sense of diminished capacity that amputees felt was reflected back in multiple ways in the camp in the diminished expectations other people had of them. Physical disability may be more closely equated with overall incapacity in populations where so much of life's daily maintenance as well as most actual occupations involve physical labor. And karmic bad fortune often seems to come in bundles, so that one instance may well be followed by another. For this reason an amputee was likely to be seen as a "bad risk" to the people around him. In any case, the loss of physical capacity was associated with a generalized loss of effectiveness and value, and amputees' overall status was clearly diminished in the eyes of the population at large. It was unusual, for example, to see an amputee in even a low-level position of authority in Site II. Amputees often excluded themselves from positions of responsibility on the assumption that even if they could do the work required, nobody would pay them the respect they were due.

This attitude was given an explicitly karmic gloss by the director of a training school for handicapped men in Site II, who told me in response to my question that no, none of his teachers were handicapped themselves, but it was better that way because they could be compassionate toward their handicapped students. The amputees, he implied, were not capable of feeling compassion for each other.

The school director was referring to that aspect of karmic theory that suggests that people who occupy different positions on a karmic hierarchy naturally behave differently. The proper Khmer response of an able-bodied person to a cripple or a beggar, and the one that the school director was referring to, is 'aanit. Often translated as "pity," 'aanit emerges in the relationship between a higher and a lower being, but it is more than just pity. It combines love with an enlightened understanding of difference, and compassion for the natural limitations under which the other must labor, because in a Buddhist universe all beings are true to their nature; they cannot help but be so. Lucian Hanks neatly illustrated this point in his classic account of merit and power in Thai society when he wrote,

Like a dog snarling to keep his bone, a lower being is more covetous than a higher one, who would generously give away his last bowl of rice. (Hanks 1976: 108)

One properly feels compassion for an amputee, not only because of the suffering involved in his injury, but because he cannot help but be what he now is: a diminished human being. There is something of this idea in the school director's comment: the amputees cannot help it, but they are not capable of behaving like able-bodied men any more. He went on to describe some of the characteristics of his students which demanded such compassion. These included hopelessness, drunkenness, violent behavior, and a tendency toward robbery and theft.

It is true that disabled people occupy a lower position on the Buddhist hierarchy of merit or virtue. Theravada doctrine is quite clear about the diminished karmic status of the crippled or handicapped,[14] and *vinaya* law limits ordination to men who are complete and whole in body, as well as being free of debt, disease, and mental deficiency.[15] I was told that even a man who was able-bodied at the time of his ordination would be forced to disrobe if he were subsequently to lose a limb. But the attitude that people exhibited most often toward the amputees seemed to come less from an assumption that they would naturally behave badly than from a wariness that came with a realistic assessment of their situation: they were former soldiers with plenty of reason for anger and despair; their employment opportunities were limited but they had just as much need as anyone else for an income; they were likely to take what they needed if there was no other way to get it. This was, in effect, a working understanding of the law of karma that suggests that people behave according to their nature. Under these circumstances wariness was a more prudent response than compassion or pity, and the only response most people felt they could afford.[16]

How amputees did behave varied of course. Some men worked hard to find ways to support their families, and learned to be resourceful with their disability. Many more found it very difficult to overcome the "karmic" handicap which involved their own and other peoples' perception of their "incapacity," however, as well as the very real limitations on their ability to work. Amputation had the effect of stigmatizing these men, and they responded differently to this stigmatization. One response was a kind of despair. "They are not convinced of courage and become people of quietness" is how this response was described to me by one Khmer who worked with amputees in Site II. Here that aspect of *samattaphiep* which involves courage, strength, and bravery is invoked. Their spirits were crushed by this "incapacitating" physical injury, the comment seems to suggest. They had lost their self-esteem; they could not face other people. They had literally lost the courage to present their face to the world.

Ashamed of their reduced circumstances and uncertain of their ability to protect themselves or their families, these men stayed close to home, leading subdued lives and avoiding social interaction. Others sought solace in the company of other amputees. Sensitive about the loss of status that attended their "reduced capacity," amputees often rejected the kind of work that was made available to them, preferring instead to gather in groups at the training schools or Handicapped Associations where they could work (or not) among people with similar experience, an equivalent social identity, and a common point of view.

These Handicapped Associations, run by the KPNLF administration, offered little more than token support for the amputees. But they did provide a meeting place for men with similar difficulties and complaints, and occasionally served to focus their anger – usually on the administration itself, which so clearly took care of its own interests but could not find resources to assist the men who had been disabled in the defense of its political cause. Amputees did not typically organize into interest groups to demand redress for their grievances. The strength – or perceived strength – of the KP leadership in Site II discouraged direct confrontation. On the few occasions when amputees did act as a group they were simply bought off with a few additional supplies. They were never able to win any substantial material or programmatic support.

More often the amputees' anger was expressed in random, anarchic ways in the camp. This kind of sporadic, unorganized violence typically occurred in the context of jealousy and/or insulted pride or "face." For Cambodians, "face" is the image one projects of oneself to the world. It is the outward expression of one's essential personality and dignity, and as such is extremely sensitive to insult. One scholar of Khmer culture has described "face" this way:

Any attempt to undermine a person's "face" is regarded as a serious injury: reproaching someone, even justifiably, in public causes that person to "lose face," or "kills" him, as the Khmer language puts it. For words kill just as much as weapons ... A Khmer is capable of ruining himself and losing those he loves in order to destroy someone who has "killed" him socially (Ponchaud n.d.: 4).

For an amputee, "face" was often one of the few personal resources left to protect. An amputee might suspect his neighbor of adultery with his wife, for example, or read her frustration at the family's new poverty as disrespect or outright insult to his position as the head of the household. Arguments developed, and these often turned violent. Amputees were not the only people to resort to violence in local disputes, nor were they unique in their sensitivity to matters of "face." Indeed, both violence and disputes involving pride or "face" were common in the daily life of the neighborhoods. But because amputees were so often ex-soldiers and therefore often had easy access to weapons, their arguments not infrequently ended with gunshots or a thrown grenade – or so was the general perception, anyway. The unpredictability of amputees' behavior and their overall association with violence were a large part of what made the population wary of them in general.

Site II as a local moral world

This general attitude of wariness toward amputees must be understood in the context of long-standing material want and overall caution or mistrust in Site II. To return to Hallowell's terminology, these amputees were part of a very specific behavioral environment, that was shaped as much by the proximity of war and a profoundly affecting shared recent history as it was by the material conditions and relations of power (and powerlessness) in the camp itself. History and long-term political and social processes were mediated through these material conditions and power relations in Site II, producing what Kleinman and Kleinman (1991, 1992) have called a "local moral world" of priorities and expectations. Within this enclosed and highly charged environment, certain kinds of behavior came to be normative and expectable, and other behaviors developed in response to those expectations. Certain key aspects of this behavioral environment, this local moral world, illuminate these normative priorities and expectations. These include long personal histories of victimization and violence, strained material and emotional resources, and an ongoing lack of security on the border.

Everyone in Site II, including amputees, received a basic food and supply ration from the UN: nobody was completely destitute. But many people found it difficult to supplement this meager subsistence ration with the limited resources available in Site II: amputees were not the only people who suffered from poverty. With the exception of a small number of administrators, military top brass, and people with business connections outside the camp, everyone struggled with their daily subsistence and guarded their resources jealously. Any extra supplies not immediately needed were converted into cash to buy clothes, medicine, or additional food, and few people had any kind of reserve. "We live hand to mouth" was a common phrase heard even among those who were living reasonably well by camp standards. There was a pervasive feeling of vulnerability in Site II, of living close to the edge of disaster.

In fact the people in Site II *were* living close to the edge of disaster, in both space and time. The war was only a few kilometers away, their fathers and husbands and sons were involved in it, and the camp itself was threatened periodically by artillery. Moreover, the circumstances in which this population was living on the border had developed directly out of the situation under Pol Pot in Cambodia, a four-year period of such deprivation, brutality and terror that "selfishness" and mistrust had become virtual instincts for survival.[17] Indeed, for many people in Site II there had been no definitive break between Pol Pot time and the present. Although daily life had become routinized over the years in Site II, the political situation changed constantly on the border and no routine could be counted upon. Violence erupted frequently inside the camp as well as outside it, and an underlying sense of insecurity remained. In this context of pervasive if diffuse insecurity many self-protective attitudes persisted from Pol Pot time, including the habit of limiting one's concern to the needs of one's family and closest friends.

In Site II amputees did not have a corner on misfortune or suffering any more than they had a corner on poverty. Everyone had been uprooted, separated from their homes and families; everyone had suffered under Pol Pot, had lost people close to them. Each family had a tale of sorrow to tell. This being the case, people kept their stories to themselves – literally, no family could afford to get involved in the troubles of another. Given the enormity of suffering within the population, this response was a realistic strategy for protection and self-preservation. As one former Site II resident put it to me, "The situation was bad for everyone in Site II. We did not think anything special about the amputees."

Within such a community of atomized suffering, what seemed "special" about the situation of many young male amputees was the extent of their degradation and shame. They were ashamed about their physical disfigurement and their loss of physical capacity; ashamed of their reduced ability to protect and provide for their families; ashamed of their diminished value, and the loss of "face" that entailed. In my interviews with amputees this profound sense of devaluation often seemed to segué into an attitude of either bravado or hopelessness in which it really did not matter to them how they behaved. It was as though nothing they could do would really diminish them further. This gave them a kind of carelessness, or lack of concern about the consequences of their actions. One man told me in response to my query about the most frightening aspect of being an amputee that nothing really frightened him anymore, as his life was not worth being frightened for. The depth of his feelings of valuelessness made reports about amputee violence easier to understand.

Power/knowledge

There is a deflating sense of inevitability about this downward slide into despair that is brought up short by an example of resistance to the seemingly pervasive "ideology" of amputation in Site II. This alternative view was brought to my attention by an amputee in Site II who read an earlier draft of this chapter and wrote to offer his comments. They demonstrate how the language of karma can serve to justify and uphold a system of power relations, which in fact persists for very pragmatic, material reasons and not because it is necessarily either morally or doctrinally "correct."

In response to my remarks about the school director cited above, this man offered a somewhat different perspective. He wrote, "Should compassion appear in action or just words and feeling?" He countered with his own experience in a training school for the handicapped, in which more than half of his teachers were able-bodied and somewhat less than half were amputees. Perhaps the able-bodied teachers just *felt* compassion toward their students when they committed themselves to teach the disabled at a low rate of remuneration, he wrote, because individually they all left to take higher-paying jobs when these became available, "while the amputee teachers, who are incapable of feeling compassion for each other, kept working to help their disabled friends, even [though] their living wages [were just] as limited."[18]

Here the writer of the letter, himself an amputee, questions the depth of his able-bodied teachers' compassion, showing how they could pursue their own advantage in the bright light of their moral rectitude, while the disabled teachers, who were not likely to have any opportunity to work for a higher wage, stayed on to labor in the shadow of their low karmic status. In fact, he wrote, amputees were now "productive of compassion" in the form of humanitarian assistance from the West, but this was often exploited by people other than the amputees who found their own ways to take advantage of it.[19]

It is hard to tell where karma leaves off and the strategic use of power takes over in these accounts. From the point of view of those who have lost limbs, a karmic explanation of their "lack of compassion" is not simply a cultural fact but rather the positioned account of someone who fails to acknowledge the prerogatives of his position. This alternative perspective is offered not to condemn the behavior of the departing teachers but to show how individual, self-serving strategies play into a much larger and more complex web of power relations which operate with and through the system of knowledge that karma represents.

This is perhaps the most important point in Foucault's analytics of power: that power and knowledge work together, mutually entailing and producing each other, and that "truth" is inseparable from the historical (material, ideological, technological) circumstances that have produced it. Foucault (1979, 1980a, 1980b) has shown us how systems of knowledge become "naturalized" and unquestionable over time; how they come to define the terms through which we understand the world and the way it works. As the central concept in a broad system of knowledge about the universe and its workings, karma has been naturalized in this way and delineates a world which facilitates the exercise of certain kinds of power. Through the working out of these power relations – between the commanders and soldiers, for example, or the teachers and their amputee students; even between husbands and wives in the camp – a world is reproduced that confirms and upholds that system of knowledge in its turn.

What the writer's comments suggest is not so much that amputees do too have compassion, but that the language of compassion obscures very real material circumstances and structures of power that limit behavioral options for all people in this situation. What kept the amputees oppressed was not their low karmic status but the poverty that deprived them of opportunities and resources with which to improve their situation, the normative attitudes and expectations that develop under these circumstances, and the priorities of war which subordinate all other concerns, producing an economy in which almost everyone had to fight for his or her own survival. "Compassion" functions as a rhetorical idiom in this context; it invokes a generalized moral concern laden with Buddhist significance, while operating strategically within a local and very particular moral world.

This is not meant to suggest that Buddhism itself was irrelevant, however, or that as a system of knowledge Buddhism had nothing of value to offer the people who live in Site II. For a doctrine that is centrally concerned with the problem of suffering, nothing could be further from the truth. Rather I would suggest that certain central concepts in Khmer Buddhism, including karma and compassion, were cast in a local light in Site II, and infused with a kind of situated significance that upheld the organization of power in this local moral world. There were other perspectives on the hierarchy of virtue in Site II, as the resistant amputee's comments demonstrate, but these perspectives received little confirmation or support in the organization of everyday life. Indeed, it was in the clear interest of existing political and economic structures to keep these alternative perspectives marginalized or out of sight.

The political construction of affect and emotion

Having taken some time to understand the "ideology" of amputation in relation to the political economy in Site II and the structure of knowledge in which it is embedded, it may be useful to step back and take a longer view of the position of amputees in the society that Site II constituted. In this regard it is noteworthy that much of the anti-social behavior of young male amputees in Site II was not found in civilian border camps associated with the Khmer Rouge that had a similar amputee population. One explanation for this difference is that such anarchic behavior was simply not tolerated in the more tightly controlled Khmer Rouge camps. Whereas in Site II, under conditions of limited resources, the attitude had developed that everyone must look out for himself in whatever way necessary – an attitude that was reinforced by the self-serving behavior of much of the KPNLF leadership – in Khmer Rouge camps one's survival and well-being depended upon one's cooperation with the highly regulated political/administrative structure, regardless of the resources available or the way one may have felt about one's injury. Social control remained the key component of the Khmer Rouge's *modus operandi*.[20]

This raises the question of what kind of behavior is *possible* in a particular political environment, and how this environment influences the experience of affect and emotion. Janis Jenkins has written movingly of the role that the state plays in the construction of emotional experience and affect in El Salvador. She writes of the state's role in constructing a "political ethos" – "the culturally standardized organization of feeling and sentiment pertaining to the social domains of power and interest" – and the emotions of those people who "dwell" in that ethos (Jenkins 1991: 140). Mary-Jo DelVecchio Good and Byron Good have described the part played by political, religious, and economic institutions in "legitimizing, organizing and promoting particular discourses on emotion" in contemporary Iran (Good, Good, and Fischer 1988: 4; Good and Good 1988: 43–63). This analysis has been more concerned with illuminating the context of certain attitudes and behaviors than with exploring the emotional experience of amputation *per se*. A full exploration of the emotions involved in that experience deserves its own article. But it is possible here to suggest some ways in which political, economic and religious aspects of the behavioral environment seemed to constrain and encourage certain modes of affect and emotion in Site II.

In the closely monitored environment of the Khmer Rouge camps, and under the understood threat of physical violence, there may have been little opportunity to act in ways that did not conform to the behavioral standards of the Khmer Rouge "organization." In Site II, however, the authority structure was looser than this. While an outward show of respect for authority was required, actual control by authority was more diffuse and ambiguous. The political leaders in Site II exercised their authority around certain key issues that affected their financial status, the smooth functioning of their administrative apparatus, and relations with their own patrons in the UN and in states that supported their political/military struggle directly. But beyond this, behavior in Site II was largely unregulated. At the most basic level administrators were not especially concerned with what their people felt or did, as long as they demonstrated support at critical moments and their behavior did not get in the way of the pursuit of larger political goals.[21] The KPNLF structure was open enough to allow for violent expressions of anger and hopelessness as long as these did not threaten the political structure itself. It could even be seen to legitimize a certain style of self-interested behavior, as its leaders were clearly not depriving themselves in the pursuit of their political goals. Indeed, taking good care of oneself was understood to be one of the privileges of power.

On the other hand conditions of material hardship, sporadic violence and a pervasive atmosphere of mistrust within the population at large inhibited the expression of certain affects and emotions which Cambodians have and do exhibit under other circumstances. In particular there has been a narrowing of the range of emotional extension or involvement with other people in Site II. Cambodians, especially rural Cambodians, tended to be suspicious or afraid of people they did not know even before the Khmer Rouge came to power. This is not an uncommon attitude among poor rural peasants whose survival depends upon minimizing the negative impact of chance on marginal economic endeavors. But the experience of Pol Pot and twelve years of uncertainty and danger on the border exacerbated those tendencies, and the experience of certain sympathetic emotions seemed to have become almost impossible in this situation. One young man expressed this very explicitly in a conversation about life under the Khmer Rouge. He said:

Everyone worked very hard. We worked like slaves under Pol Pot. Moreover, exhaustion and fatigue constricted our feelings as well as our bodies. We wanted only one thing and that was to eat. In fact, it seemed like we forgot all about helping one another. Even our beloved parents we forgot about, because we were so hungry.

The extreme physical subjection of the body under the Khmer Rouge had been replaced by physical discomfort and political subjection in Site II. But there is evidence that the kind of emotional constriction described above was still common in Site II in 1991. One Cambodian man cited earlier who had been resettled in the United States told me that when he was living in Site II he never thought about the amputees: "They had their problems, but we had ours too. Everyone was in a bad situation. We did not think anything special about the amputees. Now when I think about them I feel something terrible in my stomach. But before, I didn't feel anything." His comments suggest that even the body's visceral response to the hurt of another body may be overridden by other, prior demands on its responsiveness, as the subjected body becomes alienated by external circumstances from its own sentient experience.[22] Here in America, thinking about amputees, he felt a sympathetic response *in his body* that he was never able to feel in Site II.

In an ideal world, the proper Cambodian response to an amputee may be compassion. But for most people in Site II who had been living in a state of uncertainty on the border for almost twelve years, following four years of unspeakable brutalization under the Khmer Rouge, compassion was not simply a difficult response, it was in many cases an imprudent one, in a situation in which radical self-interest had become a virtual prerequisite for survival. The circumstances of insecurity and limited resources in Site II reinforced in this population the tendency to be extremely cautious about who they trusted, and to extend their compassion no further than to those people with whom they had the closest and most enduring of relationships. To do otherwise was to invite victimization, in a situation in which many people felt desperate enough to take advantage of any opportunity for personal gain. Sadly, this was an economy of scarcity and hoarding, in emotional as well as material terms. It was a situation in which many amputees lived up (or down) to the expectations others had of them because, given the resources available and the attitudes that prevailed, this often seemed the only real option open. Having lost so much already, there seemed very little of significance left to lose.

Conclusion

The lifeworld presented here is bleak. It is a hard world to understand, certainly from the outside, but even more so from the inside, through the experience of those who suffered through the profound events of the last twenty years in Cambodia and were living under the conditions described above. The recent history of the Cambodian people challenges the ability of any doctrine or faith to make sense of so much suffering. Buddhism offers explanations of suffering and its causes through such concepts as karma and reincarnation, and readings of the doctrine that can lift the weight of responsibility for karmic misfortune from the shoulders of individuals who suffer. But in Site II the deep need to understand one's situation was often submerged by the demands of everyday life, and these demands were played out in ways that often overcame the doctrine's ability to comfort as well. The lived significance of central Buddhist concepts were shaped for most Khmer by the overwhelming facts of powerlessness and need, in the context of a society whose structure supported a far more oppressive interpretation.

Ideally, suffering leads to reflection and enlightened understanding. But in this situation of scarce resources and overwhelming political priorities, suffering often seemed only to bring more suffering, in a downward spiral. Karma, rather than offering hope through the possibility of change, seemed to close in on amputees with a sense of immutable destiny. Their bodies betrayed this destiny which they themselves read as a sign of their degradation, just as this reading was reflected back in the behavior of the people around them. Amputees' subjective experience of their own bodies was shaped by this politically, economically, and socially constructed "truth."

However, the example of the amputee who resisted this dominant characterization shows that while a certain configuration of power supports a particular understanding and a particular *experience* of an amputated body, this understanding is neither inevitable nor even necessarily doctrinally "correct." It rests on a set of power relations which were themselves subject to political and economic pressure from outside the camp, if too infrequently from within,[23] and which are now in the process of reorganization and change, as the border Khmer begin to return to Cambodia, to a new and very different political landscape.[24] It remains to be seen whether alternative and/or resistant views of the amputees will find room for expression in the political economy of the new state of Cambodia, and the resources to support the challenge they represent to the power/knowledge structure that prevailed in Site II.

In this chapter I have tried to show how the individual experience of amputation is in fact an intersubjective experience constructed through social interaction, "a shared subjectivity constituted in the interactive space of embodied existence" as Thomas Csordas has put it (Chapter 12, this volume). I have attempted to demonstrate how embodied experience encompasses the semiotic meaning of amputation in this particular interactive space, this "behavioral environment," for amputees as well as for those people around them who escaped this particular misfortune. I have tried to illuminate the structures of power and domination that shaped the behavioral environment that Site II constitutes, and to show how the structure of knowledge that a karmic view of the universe constitutes was used to support these social structures in a mutually reinforcing but never an absolute manner. I have tried to show how the lived experience of this power/knowledge system in the isolated and resource-poor environment of Site II, and in the context of an all but unspeakable recent past, created a unique "local moral world" in which certain behaviors and responses were expectable, certain affects and emotions all but impossible. And I have tried to demonstrate through examples involving Khmer from Site II who have achieved some distance from that world that these responses and emotions are indeed "local," that even the visceral response to an amputee may change when one is responding from a differently situated position.

Finally, I would note that while I have tried to stay "close to the body" in this paper, in fact the analysis has focused more on social experience and the common reading of particular kinds of bodies in the context of Site II than on bodily experience *per se*. Much remains to be examined in the area of shared bodily experience in Site II that would extend an analysis of the social meaning of amputation even further, for example, habits of caring for the body, common physical complaints, the importance to Cambodians of the smell of things, and the practice among KPNLF soldiers of covering their bodies with magical tattoos designed to protect them from enemy bullets. Greater attention to the body's own sensibility, to its "somatic modes of attention" (Csordas 1993) in particular historical, political, and economic contexts, will yield not just a richer political economy of the body but a truly embodied political economy of meaning.

ACKNOWLEDGEMENTS

Much of the research for this chapter was carried out while I was an employee of the International Rescue Committee, through an IRC oral history project in Site II funded by the Ford Foundation. The support of both the IRC and the Ford Foundation and their permission to use data gathered through that project are gratefully acknowledged. Additional research was supported by a grant from the Department of Anthropology, Harvard University, for which I am also grateful. I would like to thank Hugh Gusterson for organizing the AAA panel for which the first draft of this chapter was written, and Byron Good, Arthur Kleinman and Stanley Tambiah for invaluable commentary on later drafts. My understanding of Buddhist doctrine and Buddhism in action was greatly enhanced by conversations with Charles Hallissey, whose insights added much to my analysis. My greatest debt is to Tom Csordas, however, whose probing observations and unfailing support brought this chapter through its many revisions. I am grateful to Tony Banbury, Kurt Bredenberg, Prey Sangha, Chimchan Sastra, and especially to Son Song Hak for their ethnographic observations and insights; I take full responsibility for any errors of interpretation. Special thanks to Bill Burke for offering the use of his powerful photographs in conjunction with this paper. Finally, I thank the people in Site II, who were patient with my clumsiness and overcame a powerful legacy of mistrust to speak with me about their difficult lives. This chapter is dedicated to them, with gratitude and respect.

NOTES

1 "Khmer Rouge" is the common name for the ultra-nationalist communist organization that took control of Cambodia in 1975, commencing a four-year reign of unspeakable brutality and terror in the name of radical agrarian revolution. The policies and practices of the Khmer Rouge have earned the labels "holocaust" and "genocide." Pol Pot presided over this bloody period, which is often called "Pol Pot time;" he remains at the head of the Khmer Rouge organization today.

2 At the time this chapter was written the border camps were in the process of being dismantled and their populations repatriated to Cambodia, as part of the 1991 Paris Agreement.

3 Technically, the word "Khmer" refers to the dominant ethnic group in Cambodia: over 90 percent of the population of Cambodia is Khmer. In ordinary parlance "Khmer" and "Cambodian" are used interchangeably. I follow that convention in this paper.

4 As a sign, the implicit reference of which is always a person, the body is particularly vulnerable to appropriation and exploitation as well. The issue of the

appropriation and exploitation of the suffering of the amputees is taken up later in the paper.

5 Part of the difficulty in understanding the experiential dimension of another person's injury lies, I believe, in the complex relationship between a subjective and a semiotic understanding of the body: the body as "experienced" vs. the body as "read." See below, note 13.

6 Foucault has been criticized for objectifying the subjects of his analyses in a way that ignores both experience and agency (see especially Terence Turner, Chapter 1 of this volume). In *Discipline and Punish* his analysis does indeed focus largely on the discipline of "docile bodies." In *The History of Sexuality, Vol. I*, subjectivities are drawn into the complex web of domination and control, but they are treated as discursive formations: they have little to do with experience *per se*. In neither study is the experience of the subjects taken to be analytically important in itself – its significance lies in the wider complex of power relations that are revealed through the analysis. But the technique of "genealogy" does not seem to me to *preclude* the consideration of experience, even though this was not a part of Foucault's own project.

7 Each faction maintained both military and civilian camps along the border, so as to keep the activities of these two populations at least nominally separate. The United Nations provided assistance to the civilian camps, since many people ended up on the border through force of circumstance rather than political conviction, and were considered victims of the armed conflict. Although I was working in a civilian camp, KPNLF soldiers came "home" to live with their families when they were not on active duty and, of course, when they had been injured in combat.

8 The theory behind providing insufficient rations was that the soldiers would be supported by the people amongst whom they were carrying out their liberation struggle. In practice it meant that they often stole food or forced villagers to provide them with rice. Needless to say, there was little political gain in this "economizing" measure of the KPNLF.

9 James Scott (1977) writes about the "social insurance" that is the basis of a client's attachment to his patron, and the cornerstone of his sense of legitimacy about the relationship: the guarantee of basic subsistence and security in exchange for his services. Scott points out that while a patron may still be able to extract his client's services if these basic obligations are not met – the client may have no alternative but to accept his patron's diminished terms – the relationship loses its moral legitimacy in the client's eyes, and will be abandoned if the client can find a more "honorable" patron.

10 The high incidence of depression in Site II, and the fact that under the Khmer Rouge many people suffered from head injury, repeated bouts of cerebral malaria, or both, are all relevant to the complaint of reduced mental capacity commonly heard in the camp as well. See Mollica et al. (1992) for a comprehensive discussion of disability in Site II.

11 This is because one has no personal connection to one's past lives; indeed, all that connects those lives with one's present life is the agency of karma. This is the basis of the Buddhist doctrine of no self. One inherits one's karma like one's genetic inheritance: it is simply a part of you from birth. Thus no one is held personally responsible for his or her karmic inheritance. One *is* responsible for what one does with that inheritance in *this* life, however, since one *can* make choices about one's actions in the present.

12 Needless to say, not all Khmer accept the limitations of their place in the social hierarchy: opposition to the corrupt, autocratic leadership of the 1960s and 1970s, which culminated in the Khmer Rouge seizing power in 1975, arose around the demand for greater access to political and economic rights and privileges (see Kiernan 1985). But even among ardent ideologues, and especially among most of the rest of the population for whom political ideology holds little real meaning, there remains a deep-seated traditional sense of the hierarchical – and funda-mentally karmic – nature of social relations.

13 Note that amputees themselves as well as the people around them "read" amputation in this way. Thus an amputee experiences his body simultaneously as both subject and object. That is, even as his subjective experience is an embodied experience, still it has a particular meaning for him; it has to be interpreted. I would submit that this is true of all embodied experience – that it is simultane-ously both a "subjective" (sensuous) and "objective" (interpreted) experience – but this is perhaps especially evident with the Khmer, who so often find under-lying meaning in a situation as it is revealed through "signs" in their everyday life. These "signs" demonstrate connections to the larger, cosmic (Buddhist) order of the universe. In this case the "sign" of the amputee's body was emblematic of a deeper moral fault, a wider pattern of degradation that extended back in time, a karmic pattern.

14 In a Buddhist universe moral status and social status are intertwined and mutually reinforcing; one cannot be reduced to the other. Karmic status entails both; it indexes both intrinsic (moral) and extrinsic (social or instrumental) value and power.

15 These regulations have to do with protecting the good name of the order of Buddhist monks, or Sangha. See Gombrich 1988: 107–9.

16 I am indebted to Prof. Charles Hallissey for this insight.

17 Pol Pot's brutal revolution attempted a radical restructuring of Cambodian society: the cities were emptied and the countryside was transformed into an enormous twenty-four-hour work camp. Between 1975 and 1979 over one million Cambodians out of a population of eight million were killed, through a combin-ation of overwork, illness, starvation and execution. See Ngor (1987) and Szymu-ziak (1986) for first-hand accounts of life under the Khmer Rouge in the 1970s.

18 I am indebted to Son Song Hak for his extensive and thoughtful comments on an earlier draft of this paper.

19 Here the issue of the commodification of the suffering of amputees and its exploitation by entrepreneurs in the world of refugee assistance comes to the fore. In contexts of extremely limited resources people do what they can with whatever is available to them, and in Site II, with its numerous, comparatively wealthy foreign relief agencies, the ability to elicit compassion and pity could be valuable indeed. Among the Khmer themselves suffering had little exchange value – the market was glutted, so to speak – but the entrepreneur who could get his product to an open market (which the Western agencies represent) or control the income generated through traffic in images of suffering was a clever and creative businessman. At the more basic level of street wisdom, every refugee knew that each new Western relief worker was a prime target for his or her best tale of woe.

20 My evidence is based on interviews with United Nations officials who worked in Khmer Rouge camps on the Thai–Cambodia border. Since I did not do extensive

research in these camps myself, my interpretation of people's behavior there is necessarily tentative. The example is offered primarily for the light it sheds on the situation in Site II.

21 This is not to say the KPNLF had no feeling for their people or that they did nothing for the social welfare of the population. Neither of these things was true. I mean rather that first, their style of leadership was not especially intrusive except around issues of strategic concern, and second that political/military priorities subordinated concerns about their peoples' individual well-being to the larger concern for the overall well-being of the guerrilla struggle.

22 Elaine Scarry (1985) writes brilliantly about the ways in which a subjected body is divided against itself, and how the world of meaning narrows around its point of resistance to that subjection, until, in the case of torture, meaning is swallowed up and obliterated in the body's own pain. Her analysis has much to add to an understanding of the subjection of bodies in Site II.

23 The maintenance and support of the Khmer resistance by interested international powers – principally the United States, China, and Thailand – was a crucial aspect of the overall power dynamic on the border. Importantly, it enabled the KPNLF to maintain a monopoly of political power in Site II, since no comparable support was available for any local opposition. This contributed to the widespread feeling in Site II that the power of the KPNLF was unassailable. But most people in Site II realized this would no longer be the case once they returned to Cambodia.

24 The October 1991 peace agreement included provisions for the repatriation of all the border Khmer in preparation for democratic elections under United Nations supervision in 1993. At the time of this writing repatriation was well under way, although political instability in Cambodia leaves the outcome of the UN-monitored peace process very much in question.

REFERENCES

Asia Watch and Physicians for Human Rights (1991) *Landmines in Cambodia: The Coward's War*. U.S.A.: Human Rights Watch and Physicians for Human Rights.

Csordas, Thomas J. (1990) Embodiment as a Paradigm for Anthropology. *Ethos* 18: 5–47.

(1993) Somatic Modes of Attention. *Cultural Anthropology* 8: 135–56.

Dreyfus, Hubert and Rabinow, Paul (1982) *Michel Foucault: Beyond Structuralism and Hermeneutics*. Chicago: University of Chicago Press.

Feldman, Allen (1991) *Formations of Violence*. Chicago: University of Chicago Press.

Foucault, Michel (1979) *Discipline and Punish: The Birth of the Prison*. New York: Vintage.

(1980a) *The History of Sexuality, Volume 1: An Introduction*. New York: Vintage.

(1980b) *Power/Knowledge: Selected Interviews and Other Writings, 1972–1977*, Colin Gordon, ed. New York: Pantheon.

Gombrich, Richard (1988) *Theravada Buddhism*. London and New York: Routledge and Kegan Paul.

Good, Byron (1994) *Medicine, Rationality and Experience: An Anthropological Perspective*. Cambridge: Cambridge University Press.

Good, Mary-Jo DelVecchio and Byron Good (1988) Ritual, the State, and the

Transformation of Emotional Discourse in Iranian Society. *Culture, Medicine and Psychiatry* 12: 43–63.

Good, Mary-Jo DelVecchio, Byron Good, and Michael Fischer (1988) Introduction: Discourse and the Study of Emotion, Illness and Healing. *Culture, Medicine and Psychiatry* 12: 1–8.

Gusterson, Hugh (1991) War and the Disappearing Body. Paper presented at the 90th Annual Meeting of the American Anthropological Association, Chicago, Illinois.

Hallowell, A. I. (1955) *Culture and Experience*. Philadelphia: University of Pennsylvania Press.

Handicap International (1990) *Thailand Program Annual Report: Physical Rehabilitation Programs for Handicapped Refugees and Thai Villagers*. Bangkok, Thailand.

Hanks, Lucien (1976) Merit and Power in Thai Society. In Clark D. Neher, ed., *Modern Thai Politics: From Village to Nation*. Cambridge, MA: Shenkman Publishing Co.

Headley, Robert K., Kylin Chhor, Lam Kheng Lim, Lim Hak Kheang, and Chen Chun (1977) Cambodia–English Dictionary, Volumes I and II. Washington DC: The Catholic University of America Press.

Jenkins, Janis (1991) The State Construction of Affect: Political Ethos and Mental Health among Salvadorean Refugees. *Culture, Medicine and Psychiatry* 15: 139–165.

Keyes, Charles F. (1983) Introduction: The Study of Popular Ideas of Karma. In Charles F. Keyes and E. Valentine Daniel, eds., *Karma: An Anthropological Inquiry*. Berkeley: University of California Press.

Kiernan, Ben (1985) *How Pol Pot Came to Power*. London and New York: Verso.

Kleinman, Arthur and Joan Kleinman (1991) Suffering and its Professional Transformation: Toward an Ethnography of Interpersonal Experience. *Culture, Medicine and Psychiatry* 15: 275–301.

(1992) Moral Transformations of Health and Suffering in Chinese Society. Paper presented at the MacArthur Foundation Sponsored Meeting on Morality and Health, Santa Fe, New Mexico, 22–24 June.

Mollica, Richard, K. Donelan, S. Tor, J. Lavelle, C. Elias, M. Frankel, D. Bennett, and R. J. Blendon (1992) Repatriation and Disability: A Community Study of Health, Mental Health, and Social Functioning of the Khmer Residents of Site Two. Working document of the Harvard Program in Refugee Trauma, the Harvard School of Public Health, and the World Federation of Mental Health.

Natanson, Maurice (1973) *Phenomenology and the Social Sciences*. Evanston, IL: Northwestern University Press.

Ngor, Haing (1987) *A Cambodian Odyssey*, with Roger Warner. New York: Macmillan Publishing Co.

Ponchaud, Rev. Francois, M.E.P. (n.d.) Approaches to the Khmer Mentality. (Unpublished manuscript.)

Scarry, Elaine (1985) *The Body in Pain: The Making and Unmaking of the World*. New York: Oxford University Press.

Scott, James (1977) Patronage or Exploitation? In Ernst Gellner and John Waterbury, eds., *Patrons and Clients in Mediterranean Societies*. London: Gerald Duckworth and Co.

Spiegelberg, Herbert (1984) *The Phenomenological Movement: A Historical Introduction*. Boston: Kluwer with Martinus Nijhoff Publishers.
Szymuziak, Molyda (1986) *The Stones Cry Out*. New York: Hill and Wang.

4   Nurturing and negligence: working on others' bodies in Fiji

Anne E. Becker

This chapter examines the impact of culturally specific notions of personhood on the experience of embodiment. More specifically, it explores how the representational uses of the body unfold in the context of the relationship between the person, the community and the body. I will first comment on the intensive cultivation of the body's surfaces and contours in some sectors of contemporary American society[1] in order to contrast it with a relative disinterest in personal cultivation among Fiji Islanders,[2] despite the obvious representational potential of the body in the latter culture. This difference is best understood as stemming from the disparity in the experience of personhood of these two cultures.

My argument begins with the premise that core cultural values are encoded in – among other things – aesthetic or moral ideals of body shape (see Ritenbaugh 1982). One can imagine that, although the body can be sculpted and adorned in a multitude of ways, the appreciation of a particular form refers to a complex set of culturally specific symbols. As is argued below, there is widespread participation in the ethos of bodily cultivation in American popular culture directed at approximating the recognized ideals of bodily perfection. Ethnographic data suggest that there is a parallel consensus regarding the aesthetics of bodily form in Fijian society. Moreover, as we might expect, the expressed admiration for a physical feature corresponds to positive sentiments about the moral and prestige symbols it evokes. What differentiates the Fijian from the American participant in popular culture, however, is not the ability to construct and admire an ideal, nor the ability to identify what values are reflected in it, but rather, the relative lack of interest and investment in attaining the ideal by cultivating, nurturing, and disciplining the body.

The personal exploitation of the representational space of bodies to construct and elaborate an image is striking in contemporary American society, where the weight, shape, and adornment of bodies reflect active participation in the ethos of cultivation as well as competitive social positioning. Comparison with Fijian ethnographic data will illuminate elements of the experience of personhood – in particular the cultural understanding

100

of the person as an individuated being as opposed to the person as a relational or interdependent member of a community – which may be predisposed toward a purely personal manipulation of body shape and weight. Cultivation of the body is pursued by whoever is deemed responsible for the body – be it a kin group, a mother, a husband, or the self who occupies it – so a personal portrayal of the self through the medium of the body depends on the formulation that control of it is the inalienable province of the individual. This exclusive right of the self to the body is protected by a cultural premium on autonomy and independence.

The use of the body to showcase the self also assumes the moral acceptability of distinguishing and aggrandizing the self, which is supported by meritocratic values in popular American culture. In fact, the portrayal of the self as an image rests heavily in the notion that the body belongs to the self and that its attributes are representative of that self's essence. Goffman (1959) wrote that a key strategy in representing the self lies in a manipulation of cultural symbols. Sontag (1979: 28) has argued that the body became popularized as a means of protecting self-image through the romanticization of tuberculosis. In discussing the notion that bodies harbor symbols of character, Turner has described the body as a "site of enormous symbolic work and symbolic production" (1984: 190). He has written that "its deformities are stigmatic and stigmatizing, while at the same time, its perfections, culturally defined, are objects of praise and admiration" (ibid.: 190).

In the last several decades, the disciplined body has emerged as a popular culture ideal in American society. Social success is contingent on the belabored construction of a particular image, which, in turn, hinges on the cultivation of "successful bodies, which have been trained, disciplined and orchestrated to enhance our personal value" (Ibid.: 111). Bodies, in this context, have become conceptualized as "strictly a work in progress" (Rodin 1992: 58) as "we labor on, in and within [them]" (Turner 1984: 190). The self operates upon the body, choosing from a repertoire of available symbols, and body shape and weight are ultimately seen as the residue of indulgence (obesity), restraint (thinness), or discipline (toned musculature). The culture validates this ethic of intensive investment in the body as a key to the projected self-image, suggesting that the goal is not necessarily to attain a particular physical feature, but rather to signal participation in the process of body work and image-making.

Bodily cultivation has arguably reached its extreme in the restrictive dieting and obsessive exercise regimens which verge toward eating-disordered behavior. Indeed the epidemiology of anorexia and bulimia nervosa reveals that behaviors associated with these syndromes are quite culturally grounded. While it is a stretch to prove that these syndromes are created by their cultural context, a conservative view at least allows that they flourish in a milieu which sanctions a dedicated concern with self-image, most notably when it seeks to emulate culturally identified aesthetic (and moral) ideals.

While a discussion of the cultural influences in the pathogenesis of eating disorders is beyond the scope of this paper, it is relevant here to note that this topic has inspired a scholarly interest in understanding our culture's apparent tendency to render bodies as objects. Feminist critique of contemporary American culture has noted the objectification of female bodies, but the exploitation of the representational capacities of the body is not restricted to women's bodies. Bodily cultivation has prospered within consumerist late Western capitalist culture which commodifies the body (see Turner 1984), and encourages work upon it to enhance its value. What remains unexplained by these perspectives is how the experience of being in a body as one aspect of being-in-the-world can potentially collapse into the more restricted experience of being in a body that is at once individuated and objectified, personal and alienated. In the following pages I will argue that an exploration of how the self is situated in and related to a body is a crucial element in understanding a culture's relative preoccupation with or disinterest in the use of the body as a medium of projecting the self, and ultimately in forging embodied experience.

The semantics of body shape

The use of the body as a malleable aesthetic medium for signaling participation in or deviation from cultural norms is well known. The virtually limitless parameters of manipulation include adornment with cosmetics and jewelry, coloring and sculpting of the integument, scarification, and reshaping of the body. The surfaces and symmetries of the human body may be understood to harmonize with local cosmology or perhaps to demarcate boundaries in the social and natural universe. In short, the ways in which a body can express meaning appear inexhaustible; this chapter will focus, however, on the aesthetic manipulation of body shape and weight and how this reflects the nature of personhood and embodied experience.

Aesthetic preferences for body shape and size are informed by core societal values. In popular American culture, these ideals appear in and are reified by the media through their association with moral and prestige stereotypes. In a well-known study in the eating disorder literature, Garner et al. (1980) argued that the shifting measurement and weight norms of *Playboy* centerfolds and beauty pageant finalists demonstrate a trend toward valuation of more slender and tubular female bodies in the 1960s and 1970s, if we accept their premise that these media figures constitute meaningful cultural icons. In a follow-up study, Wiseman et al. (1992) concluded that this trend of valuing thinness has continued up through the 1980s and created incentives to diet, presumably to emulate the ideal. There is also evidence that in subcultural groups thought to be under extraordinary pressure to be slim, there is an increased prevalence of eating disorders (Hamilton et al. 1985). Without venturing into the theoretical quicksand of whether it is valid to explore the ways in which psychiatric symptoms may be culturally expressive,[3] there is no shortage of evidence that bodies are starved, stretched, exercised, and even surgically reconstructed for the purpose of portraying the ideal self through their surfaces and contours. In short, bodies are cultivated to effect the attributes imbued with culturally relevant values which the self wishes to display.

Given the historical flux of American preferences for body shape, it seems logical that there will also be cross-cultural variability in these preferences. A study comparing female Kenyan with British subjects found that the former group perceived heavier female shapes in a significantly more positive way than did the British subjects. The positive valuation of obese female figures by the Kenyans was attributed to their association of body fat with sufficient access to food in a land of scarce resources (Furnham and Alibhai 1983). In the South Pacific, ethnographers have noted the association of social status with physical stature. For instance, Mackenzie (1985) has noted that the perceived ideal body shape in Tonga or Samoa is influenced by the relative vulnerability to food shortage, making it unlikely that prestige would be symbolized in dietary restraint.

Like Americans, Fijians express admiration for certain ideal features represented in body shapes. These ideal features, such as sturdy calves (*bodi la*), or a body which is well-formed and filled out (*jubu vina*, are also associated with particular cultural virtues – in this case, the ability to work hard and the evidence that a person has been cared for well. Moreover, the language of insults reveals a relative distaste for overly obese or thin persons and there is a clear preference for a robust form (sometimes corresponding to overweight by American standards).[4] Although these relative differences are intriguing, the most striking difference between the Fijian and the American attitude toward ideal body shapes is the Fijian's absence of interest in attaining the ideal shape as a personal goal. In contrast with the American, the Fijian does not personally cultivate his or her body to project a public image.

What remains puzzling is that despite this relatively passive stance toward their bodies, commentary on body shape – the form of one's body, whether one has lost or gained weight – is absolutely central to everyday discourse in the Fijian village. Remarks drawing attention to size and changes in others' bodies are virtually unremitting, and the language of greeting, teasing, and insults is riddled with references to weight loss or gain or unusual body size.

Although this initially appears paradoxical, it can be best understood in the context of the Fijian experience of personhood and embodiment. The fundamental orientation of the Fijian is to the community. Leenhardt wrote that the "Melanesian is attached by all the fibers of his being to the group" from which his worth is derived (1979: 94). He argued that the conceptual introduction of the body to Melanesia challenged their sociomythic view of persons as located primarily in relationships, thereby allowing "individual discrimination and a new view of the world" (ibid.: 164). This tension between the knowledge of having a body and the formulation that the "human extends beyond man's physical image" (ibid.: 26) forged the essence of Melanesian personhood as a polarity of "individuation and communion" (ibid.: 169). At the same time, the element of the person which allows social participation favors the experience of personhood as "a plenitude" (ibid.: 169) rather than as an asocial individual. Similarly, the Fijian certainly experiences his body, but his self-awareness is substantially defined by his membership and participation in a community.

What is encoded, read, and no doubt experienced in the changes and form of the Fijian body, therefore, is not the individual's finesse in tapping into core symbols of prestige, but rather, the social positioning of the person – how he has been nurtured or neglected in his social milieu. Apparently, attention to body shape and appetite in the social network is an idiom of care, nurturance and social enmeshment. Care, then, is concretized in body shape. While by comparison, the American cultivation of the body is motivated by its potential to represent personal prestige and accomplishments, the cultivation of the Fijian body records the collectivity's achievement in crafting its form.

The following discussion will illustrate the attention to individual bodies in an attempt to gauge the community's success in nurturing its members. The quality of this nurturance is measured by the enactment of duties which relate to caring for others according to complex social protocols. Specifically, the concept of care comprises several types of service which are rendered within the context of various social roles. These include: *viqwaravi*, i.e., looking after the needs and comforts of another, especially with regard to serving and attending meals; *vikawaitaki*, i.e., showing interest in and attentiveness to another's well-being; and *vilomani*, i.e., having general interest and empathy for another's problems, growth, successes, and so forth. The modes of behavior encompassed by the concept of care are most frequently and visibly evident in the procuring and sharing of food resources.[5] Finally, the cumulative effects of care are manifest in the body morphologies of the receivers. So in essence, a body is the responsibility of the feeding and caring microcommunity and consequently, its form shows the work of the community rather than of the self. What differentiates the

Fijian from the American modes of cultivation of the body is other-center-edness. In Fiji the individual body is the locus of vested communal efforts while individual efforts are directed back toward the community.

Body shape not only suggests personal abilities, but marks connection to the social network and reflects its powers to nourish. Children and guests (*vulagi*) are often targeted for extraordinary efforts in care, since their weight gain or robust forms will be credited to the caretaker's social prowess. Similarly, the collective devotion to cultivating the chief's body reflects his or her representation of the community (see Leenhardt 1979: 108). Since there is an explicit association between caring in the sense of feeding (*viqwaravi*) and weight and body shape, commentary on both *viqwaravi* and its manifestations in body morphology become parallel discourses on nurturing and negligence. The next section addresses how the performance of care is monitored in body morphology and indicates social well-being or distress.

Kana valevu and going thin: the rhetoric of care and negligence

Care-giving is central to social life in the Fijian village; it is practically expressed through the moral imperative to share food resources and materializes in the bodies that are fed. Care is also symbolically enacted in continuous rhetorical commentary guaranteeing the commitment to share food resources.

The primary cultural preoccupation with hunger and the distribution of food throughout Melanesia and Polynesia is reflected in the language of insults which has been elaborated to lambaste both those who are hungry and those who do not share food. Malinowski reported that in the Trobriand Islands, the most degrading comment to make about another is to call him "hungry" (Young 1979). Commentary which condemns those who do not share food runs parallel to this discourse. For example, Kahn (1986) describes how character judgment in Papua New Guinea directly relates to the generosity with which people share food.

Since feeding and food-sharing in the household and the community are the chief means through which social relations are conceived and maintained, Fijians are loath to be considered *kanakana lo* (i.e., one who eats secretly) and are explicit in their desire to share food. References to hunger and generosity with food can be contextualized as a commentary on care and being cared for in the social milieu. Moreover, the disparagement of hunger is a moral indictment of someone who can neither provide food nor obtain it through affiliation with a community which feeds its members. Inferences based on the appearance of the body are made explicit in the associations between thinness and deprivation (material and social) or thinness and laziness.

Sharing of food resources is axiomatic. Indeed, when a household serves food, its members are obliged to throw open their doors and windows, so that the meal is, in essence, publicly displayed and accessible. Moreover, persons who are eating or serving a meal maintain watchful eyes toward the doorway for passers-by – whether kinsmen or strangers – in order to invite them in to partake of the meal. The standard greetings, "come and eat," "there is cassava here," or "have your noonday meal here," issue forth from every household at mealtimes. Although the invitations are compulsory, they are also rhetorical. That is, while people are genuinely welcome to enter the household and eat, they usually continue on to their expected mealtime destination.

To accommodate potential guests, Fijians prepare large quantities of food – especially root crops – to allow wide margins of excess so they may confidently extend invitations to their meals. They routinely affirm that to have inadequate food supplies to feed guests or to contribute to ceremonial exchange confers disgrace. One woman explained:

[If there] is not enough for us, the members of the household are going to be ashamed that there is too little food ... we should eat, we should be well-sated (*bori vina*), never mind if there are leftovers ... if only because it should never happen that we eat just a little ... that we still feel hungry – yet the food is finished ... that would make us ashamed.

Whether the potential guest is a friend or relative stranger is immaterial. In fact, if a guest is known to be staying in the village, households will carry a food offering (*kabekabe*) to the household in which he or she is staying in order to pay their respects to the stranger.

Another central feature to the *viqwaravi* of a meal is the running verbal encouragement by hosts and servers to their guests to consume as much as possible. This rhetorical commentary during the meal complements the requisite invitations cast outside the cookhouse to passers-by. The women tending to the serving keep a vigilant watch on each plate. They urge, "*kana valevu*" ("eat a lot"), "*kana tale*" ("eat some more"), or "*e hi vo na cawa*" ("there's still more food"); alternatively, they express disappointment with the allegedly meager amount the guest has eaten, saying, "*o iko tasi kana valevu*" ("you haven't eaten much"). Oftentimes, a particularly assiduous host or hostess will insist "*kana valevu, mo urouro!*" ("eat a great deal, so that you may become fat!"). Although this is a matter of politesse, these enjoinders are often meant quite literally.

Fijians hope that their efforts to nurture their guests will be rewarded with the recognition of their *vikawaitaki* and *viqwaravi*, or care. It is assumed that these efforts are ultimately manifest in body morphology. To this end, there is extraordinary attention paid to changes in the face and body indicative of a weight gain or loss; moreover, their recurrent commentary on changes to body shape makes their assessments explicit. Hosts routinely direct their guests to eat well to avoid any possibility that their care be deemed negligent. For example, a rural village woman stated,

They'll be very, very proud if you put on weight and really look healthy before you go back to your parents; and if you go back to your parents skinny, you know, there are a lot of reasons behind it. And maybe they'll think that your parents won't want to send you back to them, and then maybe they'll think that they'll never feed you properly.

Given its explicit reference to nurturance and *vikawaitaki*, the mealtime commentary indicates concern for well-being by invoking a complex set of symbols which refer to food exchange and body morphology. The idiom is pressed to its limits in formulating the ultimate goal of the intensive care of an individual: that he should become fat. It is important to recall, however, that Fijians do not find obesity particularly appealing. Nurturing, rather than the actual cultivation of body shape, is the ostensible mandate.

Fijians monitor changes in body morphology within their network of care with great vigilance. The consequences of material and emotional deprivation or loss are thought to be manifest in bodily thinness, and use of the idiomatic "gone thin" (*a luju hara ga e lala*) evokes a certain social disconnectedness. Any evidence of weight loss is assumed to reflect a disruption in the cohesiveness of the social milieu or gross negligence on the part of the caretakers. In this respect, the affective and social position of the individual are condensed in bodily form.

The preoccupation with *macake*: the defense against an appetite disorder[6]

Given the history of nurturance and negligence codified in and read from body shape, it is not surprising that Fijians scrutinize the appetites of their charges to prevent a decline which might result in weight loss. Indeed, appetite is negotiated as the key variable in health. The loss of appetite is not only considered a herald of serious illness, but it is the pathognomonic symptom for the culturally elaborated syndrome, *macake*. *Macake* is a syndrome characterized by a variety of symptoms – a whitish coating on the tongue, sores in the mouth, a change in urine color, inflamed gums, a running nose, and fever – but its *sine qua non* is appetite disturbance. This illness is endemic in young children and quite prevalent in the adult population as well.

Descriptions of the experience most often focus on the absence of interest in eating, for example, the sensation that "one's inside doesn't want to eat" or "one just doesn't feel like eating." This lapse in appetite leads to the most worrisome feature of *macake*: its manifestation in weight loss. While bracketing the issue of whether *macake* has a correlate in Western biomedical nosology (see Good and Good 1982: 141–5),[7] the concern to identify and treat it can be understood as an elaboration of the Fijian interest in body morphology. The tandem concerns with *hunger per se*, and its opposite category, *macake*, or *lack of hunger*, underscore the ultimate interest in maintaining the body's strength and weight.

What is striking about the management of *macake* and appetite in general, is that, although sweets are directly implicated in the pathogenesis of this disease, parents rarely withhold them from their children. Instead, when children's or adults' appetites appear to wane, their caretakers immediately administer one of several traditional medication therapies called *dranu*. In fact, the regular, prophylactic dosing of these *dranu* is the community standard of care. This preference for administering medicines in lieu of withholding food is entirely consistent with the Fijian notion that feeding is equivalent to nurturing. The widespread administration of *dranu* for *macake* and also a vast range of minor ailments is a natural extension of care, by oral administration, and the cultivation of bodies as well.

Finally, the preparation and administration of *dranu* is only one way that diagnosis of *macake* is exploited as an idiom of care. Perhaps more significantly, *macake* generates the requirement that careful attention be directed to changes in appetite and body shape as possible indicators of distress in the body or distress in the social world. The practice of guarding and monitoring appetite is fundamentally integrating, since treatment of *macake* both practically and symbolically enmeshes the afflicted individual (or individual at risk) in a network of care. Hence, Fijians have not only elaborated a variety of idiomatic expressions to stimulate the appetite socially, but have also institutionalized social vigilance over the appetite by means of detection and treatment of *macake*.

Body surveillance through somatic experience

Attention to hunger is central to the Fijian ethos of care. The constant watchfulness during mealtimes to identify persons with whom to share food, the surveillance of those in the community who may be materially or emotionally deprived, and the vigilance directed toward appetite integrate individual persons into communities which can bestow care on them. However, any situation which may potentially alienate an individual is attended to by virtue of its concrete manifestations, which are not necessarily confined to the space of the body. Bodily states are not only accessible to the beholder by their visual presentation, but by their patency in additional perceptual modalities, including somatic sensation. Monitoring of bodily states is particularly intense when the integrity of the community is threatened in some way by the social isolation of an individual, such as in the case of having an illness or keeping a secret.[8] In these situations, it is notable that what the American may think of as personal experience is not encapsulated by the individual body in Fiji, but rather reverberates in other bodies and events.

Perhaps the most striking illustration of bodily experience transcending the individual occurs in the case of an undisclosed pregnancy. Within the context of the extended family, or *mataqali*, the productive and reproductive capacities of the body are theoretically appropriated toward community ends, with only tenuous individual control over allocation of bodily resources.[9] Given this imperative, the knowledge of a pregnancy must be conveyed immediately to a community. If it should be retained as a secret of the body, this embezzlement of community property will be experienced as socially disruptive.

Anecdotal accounts of undisclosed pregnancies and the havoc they bring to their communities comprise the most compelling testimony by Fijians legitimizing the social claim to individual bodies. Mishaps are often interpreted as indications of a possible violation of the moral obligation to notify a community of a pregnancy, which thereby obstruct the integration of the new member into the group. For this reason there is an inspired watchfulness for signs of pregnancy, both as bodily changes and unexplained environmental occurrences.

Women in particular are informed of and attentive to the early signs and symptoms which herald the more obvious stages of pregnancy. A young woman explained,

For Fijians, oh, the old ladies, they're smart ... they can tell when a woman is one month pregnant, or two months pregnant ... by the look of the lady, eh? the pregnant woman, eh? they can tell. They said ... in the early pregnancy, first month of pregnancy, second month, eh? they can tell by the look of the pregnant lady, she'll lose a lot of weight, eh? ... They used to say that they automatically lose weight ... they said when it comes ... when the baby inside is getting, you know, more matured or bigger, fully formed, eh, they say that that's the time when she'll gain weight again ... four months time, five months pregnant ... they'll uh, regain their weight again ... they'll get fatter that time. And also ... another sign the old ladies can really tell that she's pregnant, by you know, they'll eat a lot of unripe fruits ... like mangoes ... they won't eat the ripe one, they prefer the unripe one ... that's a sign of pregnancy.

Again, there is emphasis on weight and appetite as key variables to be monitored. Other women explained the changes as an initial weight loss noticeable in the face and prominent clavicles (*domo bale*); another woman observed that the legs, especially the calves, become thin along with the neck, and in general, "the bones will show." Later the hips and the breasts enlarge as the pregnancy develops. The complexion is also thought to become fairer.

While detected in subtleties of body shape, pregnancy manifests itself extracorporally as well in certain cases. Generally this occurs when there is a breach in the obligation promptly to inform (verbally and ceremonially) the community of a pregnancy. This non-disclosure constitutes a major moral transgression which results in a variety of untoward environmental epiphenomena. Not only is the secrecy an intolerable antisocial offense, but in many ways, the implied autonomy is unfathomable. Experience is neither private, nor individual; it is fundamentally social and diffuse.

Leenhardt suggested that the Canaque's mythic "cosmomorphic view" supports an "undifferentiated" notion of one's body from the world (1979: 20). Similarly, a Fijian woman has no choice but to reveal her pregnancy simply because her body is unable to contain the experience as a personal event. Her body divulges the pregnancy in token catastrophes: cakes not rising, chairs falling flat, and boats encountering rough seas in her presence. More threatening are the effects of her undisclosed pregnancy which manifest in other bodies. The hair she cuts may fall out, her glance may dry up the milk of a lactating mother, or her very touch may contaminate food, making it toxic to children and the frail elderly for whom she cooks. The food she prepares for a young child may even cause the child to lose weight or to "go thin."

The explanation of one woman who had not yet properly shared the news of her pregnancy illustrates her concern that she needed to ask someone else to cook for her toddler because of the danger her state posed to his health. She reasoned:

Because if I give him the food ... he is eating a lot, but he is thin now, he won't grow ... his weight will go down, even if he [eats a lot] he will just stay tiny.

In retrospect, she recalled that continuing to cook for her child had been at great risk:

[My son] went thin! You can notice it in the photo ... he looks terrible ... because I was hiding [my pregnancy] and taking care of him ... children can go thin ... or an elderly person can die.

Given the intensive concern with weight and body morphology, an undisclosed pregnancy is morally irresponsible. The guarantee of disclosure, however, is often the revelation of the secret in another's body. A

Fijian woman described her mother's experience of an "itching breast" whenever a member of the family was pregnant. The evidence was considered compelling enough to address formally the possibility of a hidden pregnancy:

She'll say, "Oh, someone's pregnant" ... There'll be signs too, in the family ... And then usually, if the signs are there, the father will call all the girls and ask, "One of you is, uh, you know, you better say it before something really bad happens."

Again, the anticipation of danger reflects a concern that the secret is toxic to the community. Its most dreaded effects are always around the interruption of nurturing activities, as illustrated in the following account by a woman of the effects of a secret pregnancy on her mother.

If [a] pregnant woman sees the breast of the lactating mother, one [breast] can become shorter, or else, she might, uh ... the milk just dries up ... It happened to me when I was a kid ... One woman ... she was pregnant ... and she saw my mother's breast, and as a result, my mother's breast dried up ... my mother had to resort to powdered milk to feed me.

In other words, intimate social knowledge of another's body is made explicit by means of its embodiment elsewhere. Verbal disclosure may be a matter of moral deliberation as an individual weighs the risk of containing private knowledge in the body, yet the information is readily available to the community since the body inevitably releases its secrets – directly or indirectly – through a second body.

The bodily-experienced surveillance of bodies is described in Csordas' discussion of "somatic modes of attention," a concept which encompasses a "culturally elaborated attention *to* and *with* the body in the immediacy of an *intersubjective* milieu" (1993). Leenhardt's notion of Melanesian personhood, which contrasts the relational, interdependent self with the relatively asocial, individuated self, helps illuminate how, although bodies are individual, embodied experience is interpersonal. When the locus of identity is simultaneously fixed in bodies and relationships, both the information conveyed by body weight and shape and embodied experiences are indicative not only of a person's situation within the flesh of a body, but his or her situation within a social plexus as well.

Since the Fijian body cannot hide its secrets nor circumscribe the person, an undisclosed pregnancy threatens the community with its powers to capsize boats, contaminate food and spoil group endeavors. In so far as it condenses multiple symbols of disruption of group integrity, the real threat it poses is of self–community alienation. This alienation is circumvented, in part, by monitoring bodies – just as body weight, shape and appetites are watched – for the evidence of social connectedness manifest in their forms.

Conclusions

In conclusion, I have compared the cultivation of bodily space by Fijians and Americans. While there is evidence that Americans labor on their bodies to exploit their representational capacities in conveying personal qualities of the self, Fijians are relatively complacent with respect to cultivating the body's space as a marker of personal attributes. Initially, this complacency seemed paradoxical. If Fijians are not concerned with manipulating self-image through the body, why did they invent and invest in rhetorical instructions to nourish – even to fatten – one's family and guests? Why, moreover, had their commentaries on the body proliferated into parallel discourses on nurturing and negligence?

Capitalist-derived values in contemporary American society encourage competitive working on the self to promote it above other selves. Post-industrial cultures invent the body by means of a calculated social representation of the self, exploiting symbols of prestige. Given what we witness in contemporary popular culture in America, we might conjecture that there is a universal aesthetic valuation of pleasing bodily forms. So, for instance, we might guess that in the Fiji Islands, where abundance of food is valued, it would be a well-fed shape that might be the operant ideal of beauty. We might further expect that Fijians are personally motivated to effect the morphologic analogs of prestige in bodily space. This is not the case, however, because in Fiji it is the capacity to be nurturant – to use food and care to potentiate social relations – which confers prestige, not the individual's achievement of any bodily aesthetic ideal. Therefore it is not the cultivation of bodies which is legitimated in Fiji, but rather the cultivation of social relationships. The success of the community in this endeavor is relative to a multiplicity of symbols which refer to the core value of *vikawaitaki* and are condensed and represented in body morphology.

Whereas the personal manipulation of the body requires an understanding of the person as autonomous from other persons, the social manipulation of bodies demands that persons are fundamentally interdependent. In his exploration of the Melanesian structure of the person, Leenhardt suggested that the "circumscription of the physical being" (in the body) encourages its individuation and objectification (1979: 164); however, he also made it clear that the Melanesian pre-Christian sociomythic legacy grounds the person firmly in social relationships. Social participation remains the essence of personhood in Fiji. Fijians may be situated in bodies, but embodied experience is not contained individually. Moreover, it is social – not personal –

identity that is concretized in anatomic space. While core cultural values may be universally manifest in aesthetic ideals of body shape, the personal exploitation of the representational capacities of the body is culturally particular. It is ultimately sanctioned (or not) by the degree to which embodied experience locates the self in an objectified body versus a nexus of social relationships.

ACKNOWLEDGEMENTS

This chapter is adapted from a paper bearing the same name presented at the American Ethnological Society Meetings in Atlanta, in April, 1990 and from *Body Image in Fiji* (Becker, forthcoming). The work is based on doctoral dissertation research conducted in the Fiji Islands 1988–9 which was supported by a Fulbright IEE Scholarship and partly by a MacArthur Foundation fellowship for joint M.D. and Ph.D. degrees. I remain grateful for the guidance and support of Professors Arthur Kleinman, Byron Good, Mary-Jo DelVecchio Good, and Leon Eisenberg in the completion of this project. I also wish to thank Dr Paul Hamburg and members of the Writing Seminar in the Department of Psychiatry, Massachusetts General Hospital for their invaluable contributions to the revision of this paper. Finally, I remain indebted to the people and government of the Republic of Fiji for so graciously allowing me to live and work in their country.

NOTES

1 From a less conservative viewpoint, observations on the interest in cultivation of bodies may be widely applicable to various sectors of the population in many Western societies. However, it is not the intent here to review the prevalence of bodily cultivation throughout these groups. Instead, my aim is to use a loosely-defined group – comprised of young and middle-aged adults in a wide variety of socioeconomic groups in the United States – which share certain practices with respect to the body for comparative purposes with my Fijian subjects, whose practices are quite different.

2 The Republic of Fiji is made up of a group of islands located in the South Pacific on the cultural and geographic interface of Melanesia and Polynesia. When I refer to Fijian cultural elements, I mean to designate those characteristic of the indigenous Melanesian people inhabiting the islands. These ethnic Fijians have remained relatively culturally segregated from the European and Indian settlers (who now comprise more than one-half of the nation's population) and have retained major elements of their traditional subsistence agriculture economy and political system of hierarchical chiefs. In short, Fijian late twentieth-century culture, while located within a westernizing political national economy, is quite distinct from the European and Indian cultural presence in the island group.

3 Whether dieting and exercise lie on a continuum with anorexia and bulimia nervosa remains controversial (see Garner et al. 1984; Katzman and Wolchik 1984). It is intriguing to consider that culturally validated body cultivation may border on pathological behavior. It also raises a question: does cultural permission to starve and abuse one's body rely on (and also reinforce) an experience of the

body as an alienated object? Moreover, it is tempting to speculate that, in Fiji, where the prevalence of eating disorders is relatively low (Prosper Abusah, letter to the author, 12 October 1992), such behavior might be discouraged by the culture's prevailing resistance to any appropriation of the body as a purely personal forum for communicating distress.

4 I do not mean to exaggerate the differences in body-shape preferences between Fijians and Americans; this runs the risk of caricaturing them in misleading ways. Indeed, my survey data (see Becker forthcoming) showed some remarkable similarities with respect to attitudes toward various female body shapes. As I argue in this chapter, it is not necessary to emphasize polarities in body-shape preferences in order for Fijians to provide a foil for Americans with respect to bodily cultivation.

5 Ravuvu provides a detailed description and analysis of a variety of Fijian ritual food prestations in *The Fijian Ethos* (1987).

6 This appetite disorder is distinguished here from the Western nosologic category of eating disorders. As I shall clarify, these appetite disturbances are culturally conceived of as physiological, not psychiatric, illness.

7 There are syndromes nearly identical to *macake* in other areas of the South Pacific, for instance *fefie* in Tonga and *popome* in the Solomon Islands. Some biomedically inclined persons have suggested that the whitish covering on the tongue and oral lesions associated with high consumption of sweets reflects oral thrush (candidiasis), although this hypothesis has not been clinically investigated to my knowledge. I would argue that while the symptoms of these two illnesses may overlap, *macake* comprises a far broader category of complaints which focus on appetite disturbance as the cardinal symptom as opposed to the physical signs I have described.

8 This attention to bodily states as evidence of social disruption is also striking when illness simultaneously threatens body and community integrity. While perceived somatically by the afflicted individual, illness is often experienced by the community as a socially dystonic event which compels reflexive inquiry about moral transgressions which precipitated the somatic disruption.

9 The American notion of the self within the body stems in part from a model that bodily activity is generated as a volitional neuromuscular event, and that somatic sensations associated with embodied experience are bound to the individual body. The Fijian understanding of the self, however, allows that a person's volitional command of the body may occasionally be incapacitated by trespassing spirits (which may appropriate bodies temporarily for community business) and that a person's desires may be subordinated to what the community mandates.

REFERENCES

Becker, Anne (forthcoming) *Body Image in Fiji*. Philadelphia: University of Pennsylvania Press.
Csordas, Thomas (1993) Somatic Modes of Attention. *Cultural Anthropology* 8: 135–56.
Furnham, Adrian and Alibhai, Naznin (1983) Cross-cultural differences in the perception of female body shapes. *Psychological Medicine* 13: 829–37.
Garfinkel, Paul E. and Garner, David M. (1983) The Multidetermined Nature of

Anorexia Nervosa. In Padraig L. Darby, Paul E. Garfinkel, *Anorexia Nervosa: Recent Developments in Research*. New York: Alan R. Liss, Inc. pp. 3–14.

Garner, David, Paul E. Garfinkel, Donald Schwartz, and Michael Thompson (1980) Cultural Expectations of Thinness in Women. *Psychological Reports*. 47: 483–91.

(1984) Comparison Between Weight-Preoccupied Women and Anorexia Nervosa. *Psychosomatic Medicine* 46(3) (May/June): 255–66).

Goffman, Erving (1959) *The Presentation of Self in Everyday Life*. New York: Doubleday Anchor Books.

Good, Byron and Good, Mary-Jo DelVecchio (1982) Toward a Meaning-Centered Analysis of Popular Illness Categories: "Fright Illness" and "Heart Distress" in Iran. In A. J. Marsella and G. M. White, ed., *Cultural Conceptions of Mental Health and Therapy*. Dordrecht, Holland: D. Reidel Publishing Company.

Hamilton, Linda, J. Brooks-Gunn, and Michelle P. Warren (1985) Sociocultural Influences on Eating Disorders in Professional Female Ballet Dancers. *International Journal of Eating Disorders* 4(4): 465–77.

Kahn, Miriam (1986) *Always Hungry, Never Greedy*. Cambridge: Cambridge University Press.

Katzman, Melanie and Wolchik, Sharlene (1984) Bulimia and Binge Eating in College Women: A Comparison of Personality and Behavioral Characteristics. *Journal of Consulting and Clinical Psychology* 52(3): 423–8.

Leenhardt, Maurice (1979) *Do Kamo*. Chicago: University of Chicago Press.

Mackenzie, Margaret (1985) The Pursuit of Slenderness and Addiction to Self-Control. In Jean Weininger and George M. Briggs, eds., *Nutrition Update, Volume II*. John Wiley and Sons.

Ravuvu, Asesela (1987) *The Fijian Ethos*. Suva, Fiji: Institute of Pacific Studies, University of the South Pacific.

Ritenbaugh, Cheryl (1982) Obesity as a Culture Bound Syndrome. *Culture, Medicine and Psychiatry* 6(4) (December): 287–94.

Rodin, Judith (1992) Body Mania. *Psychology Today* 25(1) (Jan/Feb.: 56–60.

Sontag, Susan (1979) *Illness as Metaphor*. New York: Vintage.

Turner, Bryan (1984) *The Body and Society*. Oxford: Basil Blackwell.

Wiseman, Clare V., James J. Gray, James E. Mosimann, and Anthony H. Ahrens (1992). *International Journal of Eating Disorders* 11(1): 85–9.

Young, Michael (1979) *Ethnography of Malinowski*. London: Routledge and Kegan Paul.

5 The silenced body – the expressive *Leib*: on the dialectic of mind and life in Chinese cathartic healing

Thomas Ots

In medical anthropological discourse, it has become practical to speak of culturally constructed disorders, culturally inscribed, shaped or destined bodies, even of culturally constructed bodies. If we define the relation between culture and the body syntactically, simply in terms of a subject–object relation, then culture comes to share the space of the subject, with "body" as its unproblematical object. Keeping in mind the role of the body as culture-generative also, I shall try to problematize the subject–object relation of culture and body by introducing a second concept to share the body's space. What I want to show is that the process connecting culture and body emerges continually as the structural objectification of lived bodily experience, itself assuming, through reification/representation, a negating opposition to life-experience.

A model relation of mind and body-as-*Leib*

Body-as-Leib

German language knows of two different terms to refer to the body: *Körper* and *Leib*. *Körper* takes its root in the Latin *corpus* and refers to the structural aspects of the body. It is the objectified body (somebody else's *Körper*), and also the dead body or corpse. In contrast, the term *Leib* refers to the living body, to my body with feelings, sensations, perceptions, and emotions. There are many commonly used phrases where *Körper* could not be substituted for *Leib*, e.g., *Leibspeise* (one's favorite food [that makes me feel well]), *leibliches Wohl* (personal well-being), *Leibarzt* (a personal physician who cares for me as an individual).

Leib bears a non-coincidental resemblance to the English term "life." Both terms derived from common roots *leip*, *loip*, or *lip*. This became *lif* in Old English, Old Frisian, Old Saxon, and Swedish, *libo* in Old Teutonic, *lib* in Old High German, *lip* in Middle High German, and *Leib* in Modern German. In this original sense, these terms referred to "life," "person," and "self," namely, a person-self as constituted by the quality of being a-life

116

(being a-*Leib*). In contrast, the term *Körper* views the person as a vessel/ container to be filled with the spirit or the soul. The English equivalent body has its roots in *bodig*, in present German *Bottich* which means barrel. The change of meaning from *lif* to body in the English language has an analogue in Chinese. 2,000 years ago, e.g., in the writings of Zhuangzi, the Chinese concept *shen* meant "self" as undifferentiated by the "body." Only later, due to the increasing dichotomization between mind–heart and bodily emotions, was *shen* reduced to the meaning of "body."

Leib has no semantic equivalent in modern English and its meaning is therefore difficult to render in translation. Most English translations of the works of German phenomenologists (as well as those of the French phenomenologist Merleau-Ponty who kept the use of *Leib*) use "body" for both *Leib* and *Körper*. This creates a potential epistemological problem by hiding a historical passage from *Leib* to body (*Körper*). When using "embodiment" to do conceptual justice to the phenomenon *Leib-Körper*, we may again describe the process of something else taking possession of the body, as becoming embodied. This "something else" can either be culture or the mind. The term embodiment may perpetuate the clean subject–object dichotomy of mind and body rather than helping to collapse it. René Devisch[1] suggested that we use instead the term "bodiliness" which has the advantage of not implying the directionality of this "something else (see also the usage by Terence Turner, Chapter 1 of this volume)." However, it is a frail advantage: bodiliness is too static, as it omits both the life and the motions of the body. Drew Leder (1984) has expressed a preference for the concept "lived body" which comes closer to the *Leib*. But compound terms are not convincing in constructing a paradigm, and more often than not reflect an epistemological embarrassment at its originary site.

What is the difference then in speaking of *Leib* rather than of "body"? The (untranslated) term *Leib* opens up a hidden dimension of the body, that of the body as individual. It invites us to keep an awareness of life in which perceptions, feelings, emotions and the evolving thoughts and considerations are all intimately grounded; it is through *my Leib* that I am inserted into this world; I am *Leib-in-the-world*. When thinking and talking of the *Leib*, rather than talking of the body, I virtually feel my difference. Hermann Schmitz,[2] Germany's most eminent current philosopher of the *Leib*, holds that subjectivity is intimately bounded to affective affliction. That which is communicated has to touch or hit me before I can make it "my" thing. Being touched means that I am affected, moved, or im-pressed. Someone else's grief, for example, becomes mine only if (in contrast to culturally learned compassion) it presses down on my chest or hits me in other perceivable ways. In German, the word *Gefühl* makes clear that emotions are *leib*ly felt.

Mind

If life, thinking, and culture are grounded in *leib*ly experiences, what role does mind play? Using the definition of mind given by Max Scheler (1962 [1928]), we can see the mind (*Geist*) as that principle which stands facing life. Here mind is characterized by an existential (total) independence from all parameters of organic life, be this instinct or intelligence. Given this arrangement, man is the only being capable of opposing himself (in mind), negating himself, setting limits to his own life forces, and to making himself the object of pure (disembodied) thought. Man can say *no* to himself and his needs; exercising his mind's pure will, man can be an "ascetic of life." Freeing him from life, the mind objectifies his world. Everything may become object. The mind is neither subject to internal drives nor worldly forces, nor is mind being-in-this-world, the mind is "*Umwelt*-free."

The controlling and regulating action of the mind separates the person from her *leib*ly drives and instincts. This "sublimation" then constitutes a gain of freedom and opportunity, which Scheler calls the "ideation of life." Thus, Scheler does not view the opposition of mind and life as irreconcilable, but sees both bound together on the same conceptual plain. However, he fails to elaborate on this dialectic.

I understand the relation between mind and life as more problematical: it is life, and only life, that is capable of making mind potentially active. Ontogenetically as well as phylogenetically, the sequence of "enlivenment of the mind" (Scheler 1962 [1928]) is difficult to trace. The mind as expressive of cultural structure and social relations imposes innumerable constraints and limits on the *Leib*. Mary Douglas (1982 [1970]) pointed out that bodily experiences are silenced where the social dimension is highly structured. I will take up this point in my analysis of the recent phenomenon of cathartic *qigong* in the People's Republic of China.

The Chinese "heart–mind controls emotion" model

In Chinese culture, a variety of concepts of the body are used in theoretical discourse. The concept I wish to refer to here is one which has its roots in the two most prominent philosophical schools, Confucianism and Taoism. Benjamin Schwartz (1985: 173) has made explicit "the emergence of a common discourse" with respect to some key concepts of body in Chinese thought, such as those of *qi*, nature, and heart. This discursive practice is best documented in traditional Chinese medicine (TCM).

In traditional Chinese medicine, the mind is placed within the heart (*xin*). The heart is understood as the supreme viscera because it is perceived as the seat of cognition and of virtue (*de*). The heart–mind is the grounding space for all aspects of bodily as well as social well-being. Any harmony can be threatened by the activities of the Seven Emotions (*qi qing*).[3] In TCM, excessive emotions are understood as one of three classes of pathogenetic factors. Emotions literally express themselves through bodily changes: they are utterings of the body. According to the theory of systematic correspondences (*wu xing xueshuo*) these emotions, when in excess, hurt their corresponding viscera as well as the encompassing heart–mind. Emotional behavior, therefore, is heavily stigmatized. This contrasts the Chinese understanding from the Western notion of psyche. Michelle Rosaldo (1984: 143) wrote that "emotions are thoughts, somehow 'felt' in flushes, pulses, 'movements' of our livers, minds, hearts, stomachs, skin. They are embodied thoughts." From a Chinese perspective, this statement does not make sense, because thought and emotions are strictly differentiated. The Chinese view comes closer to Scheler's definition of the mind as the "ascetic of life." There exists no equivalent of our construct of psyche in Chinese language, nor of our concept of psychosomatics. The Chinese have translated the Western term psychosomatic medicine as *xinshen yixue*, literally "heart–body medicine." Thus, the argument that traditional Chinese medicine does not share the Western dichotomy of psyche and soma, and for this reason, that its outlook has always been holistic, is truly a cross-cultural misunderstanding: we simply employ different mind–body models.

TCM's notion of heart–mind versus emotions is based on classical Taoist thought and its inside–outside symbolism, e.g., in the Taoist classic Zhuangzi, emotions and passions are external worldly affairs that hurt man's interior. Exemplary man cultivates his interior by staying calm and becoming unattached: he visualizes his interior, i.e. visualizes deities or lights inside his body. His mind travels within his body using fixed routes, the *jingluo* conduits. The Taoist adept differs considerably from the Christian monk in his attitude towards the physical body. Whereas the Christian monk disregards his sinful flesh altogether, the Taoist seeks immortality on the basis of a strong, healthy, (well-kept, cultivated) i.e., emotion-free body. The physical body becomes the basic working material of the Taoist, it is his *materia prima* (Robinet 1989: 160). Should he achieve the purest clarity of heart–mind, yet his body fall ill, he could never reach the stage of *shenxian*, thus he could not become a Spirit-Immortal.

In conclusion, I submit that traditional and current Chinese body concepts differ considerably from Western ones in their specific understanding of the lived body. Yet they do mimic the basic patterns of subject–object relations. In Chinese thought, this becomes the relation of dominance and control of the cognitive heart–mind over the stigmatized emotional *Leib*, the body in motion and turmoil.

A brief account of the recent history of *qigong*

Qigong is most often referred to as "breathing therapy."[4] This rendering of *qigong* is rather misleading for it does not take into account the metaphorical nature of the concept *qi*. *Qi* denotes the air we inhale as well as bodily forces which might be interpreted in an energetic or emotional sense. The concept *qi* bears some resemblance to the ancient Greek concept *pneuma* as well as to the Indian *prana*: the air we inhale becomes life-giving force, a cosmic breath. In common styles of *qigong* exercises, the adept relies on the mind's will power (*yi*) to guide his *qi* through his body via certain routes, the *jingluo* conduits. These "endeavors of *qi*" (a most literal translation of *qigong*) may or may not be accompanied by certain external movements.

Qigong is in fact part of an array of techniques of health preservation and of exercises prolonging life (*yangsheng*) which are intimately related to ancient Taoist practices. Over many centuries, these practices were confined to Taoist monasteries or passed on only as family traditions. In the fifties and sixties, however, the first two Chinese *qigong* hospitals were opened; various styles of *qigong* then came into use for therapeutical purposes. These forms were called *jinggong* (quiet *qigong* with little external movements) or *neigong* (internal *qigong*). During the Great Proletarian Cultural Revolution, *qigong* was discredited as superstition and black magic; *qigong* practice was suppressed, the clinics were closed or converted to other purposes.

The year 1980 marked the starting point of a unique phenomenon in recent Chinese history, later to be referred to as "*qigong* fever," or "*qigong* craze" (*qigong re, qigong rechao* [Qi 1988]). Within a short period of time, parks and other public places were filled with people practicing *qigong*. Chinese estimates variably claim that, during the eighties, there existed between 300 and 3,600 different forms of *qigong*, most of them new ones. Supposedly, there were 60 million followers of *qigong*. In 1990, the Hong Kong-based magazine *Contemporary* claimed that rapidly growing *qigong* organizations had become the strongest power besides the Communist Party.[5] Between 1980 and 1987, six *qigong* journals appeared and, by 1988, approximately 5,000 *qigong*-related articles had been published.

One aspect of this *qigong* craze is quite remarkable. In 1980, new forms like "crane-*qigong*" (*hexuangzhuang qigong*) and the "spontaneous form of the five animal movement *qigong*" (*zifa wuqinxi donggong*) came into existence. In crane-*qigong*, the adepts start out with five sets of slow, gentle, and regulated movements which purport to mimic the movements of the crane. These movements are accompanied by certain breathing techniques and by the guided flow of *qi* throughout the body. The five sets of movements culminate in a sixth and final one called the "standing pole" (*zhanzhuanggong*).

5.1 Practitioners of Crane-*qigong* in the final stage of catharsis.

5.2   The master of the group ejects *qi* on a practitioner in order to support her catharsis.

The practitioner has attained the highest stage of harmony and calmness. At this stage, however, practitioners give in to various bodily movements and also sounds. They may shake their hands or limbs, the head or even the whole body; they may jump up and down, trample the ground, move in different directions, float around in a dance-like fashion; they may massage or beat certain body parts or fall down to the ground; they may shout, scream, laugh or cry, touch or embrace others, etc. In 1984, when I first observed practitioners of crane-*qigong*, I was surprised by this 'non-Chinese' behavior. Many of the practitioners were in a state of limited control of their actions; others seemingly had entered varying states of trance, ecstasy, or emotional catharsis.[6]

These new forms of *qigong* were called *zifa donggong* (*qigong* of spontaneous movements). Although only a few of these forms came into existence, and although they were outnumbered by hundreds of quiet-*qigong* forms, these few were the ones to gain widest popularity. In 1983, it was believed that crane-*qigong* had attracted some 4.5 million followers (Wang 1985).

In the early eighties, the majority of TCM hospitals were not using *qigong* as a therapy as many now do. In 1984, while I was working in the outpatient department of acupuncture at Nanjing College of TCM teaching hospital, a decrease in the number of our patients revealed that many patients were no longer seeking relief through acupuncture or herbal therapy, but instead, had joined *qigong* sessions organized and conducted by medical laymen. *Qigong* was becoming a serious alternative to the established therapeutic tools of TCM. *Qigong* was purported to offer effective cures for every kind of disorder. Some healers like the Peking resident Guo Lin claimed that their practice would cure cancer.

It is noteworthy that over a period of ten years, none of the six major journals on *qigong* offered an account of how the "spontaneous movements" had come into existence. When I interviewed different healers on this matter, I soon realized that this was a rather controversial topic. Indeed, as early as 1980, voices had been raised against spontaneous-*qigong*, and it was claimed that it might induce serious health disturbances (*piancha*) including stroke, myocardial infarction and mental disorders. Some writers expressed the opinion that the movements in themselves were expressions of brain and mental defects. For almost a decade, spontaneous-*qigong* became the subject of heated debates. The furor this created helps to shed some light on the relation between Chinese culture and *leib*ly expression.

Political discourse and the silenced body

Here I draw on a few examples of the wealth of arguments written on spontaneous-*qigong*. If we begin with the initial issue (1980) of the first *qigong* journal (*Qigong zazhi*), we find a decisive address by Lü Bingkui (1980), then chairman of the TCM section at the Chinese Ministry of Health. In it, Lü produced eleven statements that can be interpreted as the Chinese government's view of the future development of *qigong*. The most important ones for my purposes are:

1 *qigong* is part of TCM and thus part of China's cultural heritage;
2 *qigong* helps to preserve health and prolong life;
3 *qigong* can be practiced only via the control of one's mind over the body, especially the emotions;
4 *qigong* practice must follow the paradigms of quietness, relaxation and internal harmony; the emotions must reveal stability, i.e., calmness.
5 *qigong* prerequires moral self-cultivation, but it also leads to moral self-cultivation.

The statements of this address created context for the future development of *qigong*. They attempted to bind *qigong* to China's traditional ways and the values of quietness, relaxation and harmony. These values, in turn, were presented as attainable through control by the cognitive mind over the emotional body. But adepts of spontaneous-*qigong* challenged these norms. They were loud, moving wildly, and obviously not in control of their bodies and emotions. The arguments against spontaneous-*qigong* became subsumed in the following statement, which was quoted over and over again in the journals:

Big movement is not as good as small movement, small movement is not as good as no movement, no movement is not as good as quietness. (*Da dong buru xiao dong, xiao dong buru bu dong, bu dong buru yi jing.*)

"Big movement" (*da dong*) became the leading metaphor of stigmatized behavior in *qigong*.

During the same ten-year period hundreds of articles appeared in defense of spontaneous-*qigong*. The most important issue became the very nature of the spontaneous movements. Most writers placed the issue within the traditional cultural framework of the theory of *qi*. But the resulting explanations were often contradictory. Some healers said: "As soon as the *qi* flows there will be movements"; others, referring to the classical *yin–yang* theory, explained: "extreme quietness will change into motion"; and there were those who, like Zhao Jinxiang, founder of crane-*qigong*, referred to one of TCM's pathogenetic theories which explains disease as caused by blockages of *qi* conduits. Zhao claimed that "movements will develop as soon as internal *qi* meets a blockage within the *qi* conduits" (all quotes: Wang 1985: 44).

Over the years, many different explanatory frameworks were introduced to explain either the character or essence of "spontaneous movements." Some frameworks relied mainly on arguments borrowed from modern Western physiology and neurophysiology. In a very few cases writers referred to Western psychology, and, among hundreds of offered explanations there was one short article whose three authors referred to Sigmund Freud's theory of the unconscious and interpreted the "spontaneous movement" as an expression of suppressed emotions (Ma et al. 1988). Another author, actually favoring a purely physiological approach, at the end of his article advanced the idea that "spontaneous movements" could possibly be explained as an opening up of previous constrained (depressed) emotional states (Li 1983; see also He Zongyu 1985: 11).

My cursory review of classical as well as a variety of modern approaches given in *qigong* journals should point toward a common feature: there are no attempts to account for the role of emotions. With the two exceptions stated, all modern approaches define human action solely as the function of the mind.

Emotional discourse and the expressive *Leib*

Come out, ye wonderful tears,
you relieve my heart and make me happy;
fountainwater washes away all the misery
hidden deep inside my heart.

On various occasions between 1984 and 1991, I joined in practice and did field research on a crane-*qigong* group in Nanjing (Ots 1990a). I had observed groups in Beijing before, but this one was surely the most committed. The group consisted of about a hundred persons, was largely informal and open to everybody. There was no fee taken for services. Most of the practitioners had joined the group because of some health problem. The group was headed by a healer who gave the commands for the movements, and practitioners just joined in. They met at a quarter past five in the morning in a public park, and dispersed about an hour and a half later. During the sixth stage, the leader and some temporary assistants helped and aided those who were in the stage of trance or ecstasy. After the session, the practitioners chatted for a while and exchanged views. Over time, about one-fourth of the practitioners remained with the group and joined it regularly almost every morning. Some of them had become healers themselves, others just tranced. But most of the practitioners remained with the group for a few months only. They left it when they felt better.

I had no difficulties entering. The leader only knew that I was working at the local TCM hospital. He just asked me to join when he saw me watching the group. I thus entered the group as a follower of *qigong* rather than as an anthropologist. After a short time, and also guarded by darkness, I developed spontaneous movements and tranced.

After a few weeks of practice, I began interviewing patients and healers about different aspects of the spontaneous movements. I was most interested in the phenomenology of their individual perceptions. What actually did they feel in their trance-like state? What did the different movements express? However, their answers were disappointing. They usually related that they had been suffering from specific diseases, and that *qigong* practice had cured their disease. The movements just happened. They also referred to the classical theories of *yin–yang* and the flow or blockage of *qi*. Interviewees described that they felt their *qi* flow quickly or smoothly, that they felt how it left the normal conduits and rose to their heads or filled their thorax or abdomen. When I interviewed the leader of the group and some of his assistants, they too resorted to the well-trodden explanatory models of TCM. The question thus remained, why do followers of spontaneous-*qigong* behave openly in a culturally stigmatized way, yet give culturally acceptable explanations for their behavior?

During the time I spent in China, I became close to many practitioners as well as to some *qigong* instructors. I did not mind annoying them and repeatedly enquired about their perceptions. With our developing intimacy, they stopped referring to the previously mentioned theoretical models, and eventually talked about their feelings and their sufferings, and about the manner in which they believed spontaneous-*qigong* had helped them to alleviate their individual burdens. They progressively revealed a thick, rich and complex emotional world which had been previously hidden from me. In most cases I needed six to eight interviews, or rather informal chats, to reach this hidden stratum. When I talked to the leader of the group about this, he smiled and agreed: "Yes, spontaneous-*qigong* helps to free our suppressed emotions." "But why not write something about this in the leaflet that you distribute to the patients?" I asked. "How could I?" was the answer, "If you want to be taken seriously you have to follow a certain style of argumentation."

While giving courses on crane-*qigong* in places away from home, the master asked his disciples to write to him about their experiences. He received countless letters which revealed an intense emotional world unheard of in the *qigong* journals. In most case-reports carried by those journals, the accounts of *qigong* practitioners told of suffering by naming a certain disease like stomach ulcer, hypertension, and the like. In the private letters, patients used a less labeling code; they gave detailed descriptions of their suffering by referring to perceived "symptoms." Furthermore, they verbalized their psychosocial misery, feelings of depression, unhappiness, hopelessness, anger, irascibility, sadness, terror, or fright. Some talked about their feelings of shame and guilt, or of how they had been driven to despair, how their life circumstances had made them cry, and how they felt at odds with the world. Some of them had contemplated suicide. They wrote of how they were forced to swallow their emotions and how this had changed their *leib*ly awareness, had made them choke, had oppressed their chest and bloated their abdomen, etc. The Cultural Revolution, or marital and job-related problems were most often given as reasons for their misery.

The letters also provided detailed descriptions of the different move-ments, the connection between them, and the perceived emotional changes. Only a minority of these patients alluded to the concept of *qi*. With this material, I was for the first time facing expressions of happiness which were totally absent in all the writings of the *qigong* journals:

At the beginning of practicing, my shoulders turned warm, and then also the neck and the back. After I had practiced for a certain time and up to a certain standard, I felt a flow of warm *qi* which expanded from point *dantian* [below the navel] to acupoint *huiyin* [at the anus], and later it flowed down to acupoint *yongquan* [at the sole of the feet]. But when I get into the state of spontaneous movements, then my whole body feels warm. At some parts, it feels like burning fire. When I lie down on the ground, it feels like lying in warm sand. A current of warm water rushes through my body. In fact, the whole body is filled with warm water, and it flows to and fro, to and fro, to and fro. It is such a wonderful and happy feeling as one usually can not imagine.

As long as this patient practiced *qigong* according to the norms of guided flow of *qi*, i.e., as long as he kept control over his body, his perceptions remained close to the traditional images. His *qi* flowed from one acupoint to another according to the known conduits of the *jingluo* system. But as he reached the state of spontaneous movements, he no longer experienced his bodily motion as a regulated flow of *qi*. He gave way to a feeling of his unstructured *Leib* which was intimately connected with happiness.

A woman described how she developed spontaneous movements after four days of practice, and how bodily movements made her become aware of her emotional state:

I felt a stream of *qi* leaving the earth, lift me upwards and then turn me around in different directions. I had trained for three days, thus I felt self-assured and did not go against it. Suddenly, a force pushed me in the back. I stumbled forward and fell down on my knees. Now I became frightened. I wanted to finish the session, but before I could get up, another explosion of energy hit my front and pushed me backwards. I fell to the ground, and then this energy just whirled me around and around. Again and again, I tried to stop it, but I just didn't succeed. Then, for the first time in all these years, I became aware of all my sadness and shock. I started crying. What a relief!

A short paragraph in another letter demonstrates the relation between everyday mindful action, growing *leibly* awareness of one's suffering, and the cathartic function of the trance-like state of spontaneous-*qigong*. A middle-aged woman, sent to the countryside (*xia fang*) during the Cultural Revolution and later not allowed to return to her native Shanghai, wrote a long letter to the healer. She gives a very emotional account of her life-history which brought her close to suicide. At the end of her letter, she writes that crane-*qigong* had helped her to cry. But she knows that she still needs a bit of time before she is able to "spit out all that anger-*qi*" that she had "swallowed over all these years." She writes:

Today, when contemplating all of this, I cried all day long. But you cannot compare this crying with my experiences when being in the state of spontaneous movements. Then, I do not think of any of these bad things. They just come up and pour out of me.

She remains, however, undecided whether her *leib*'s revolt against cultural norms is good or bad. She continues:

I would like to know whether, in everyday life, I worry too much? Is that the reason why all of this comes up when I am engaged in spontaneous movements? Should I control my thoughts more? My husband says that, while in *qigong*, I am too impulsive and that I should quit practicing it.

A 40-year-old man who suffers from ulcerative colitis and cancer of the testes had rather unsuccessfully tested different forms of quiet-*qigong* before he began to practice crane-*qigong*. After ten days of practice, he experiences spontaneous movements. He tries to control them by guided breathing, and also by pressing certain acupoints. After another three weeks, spontaneous movements and big laughter develop during his practice of quiet-*qigong*. Again, he tries to control them through guided action, but he does not succeed. He tells the instructor:

My movements no longer follow my thought (*yinian*); instead, my thought and images develop out of the movements and are guided by them.

In another letter a worker in a shoe-factory writes a moving poem about sociocultural pressures and the freedom of emotional expression within the context of *qigong* practice.

The tears of a *qigong*-friend

Teardrops run down the cheek of a *qigong* friend
just like morning dew drips off a lotus leaf,
or like a well, like flowing water from the earth.
This is really a wonderful, a god-like *qigong*
that understands to express all the hurt feelings and bitterness of your
    heart.
What is it, that makes you open up?
Is it worries or is it hopelessness,
is it suffering from love,
or the calls of your old mother?
Or are you alone at a distant place, yet in thoughts with your beloved ones?

The girl over there is quietly sobbing,
her inside is in emotion,[7]
yet, she is still under control.
She is totally excited but nevertheless knowing.
You will not find a more ideal chance than this one,
where *qigong* is responsible for all the expressions of your hurt feelings.
Other people will not mock you,
and nobody will poke his nose into your problems.

At a normal day, you have to hide all your tears deep inside your heart;
crying requires a quiet and hidden corner.
But today . . .

Come out, ye wonderful tears,
you relieve my heart and make me happy,
Fountainwater washes away all the misery
hidden deep inside my heart.

There is little here of cultural specificity which would make the poem inaccessible to us: it closes in on our hearts in a language that we understand, a language expressing the immediacy of the acting *Leib*. This poem moves the heart, takes to it, makes us emphatically feel the plea of this woman. As Linck (1991) noted, in private discourse in China, and also in poetry, the heart is not understood in its philosophically derived meaning of the cognitive mind but is rather experienced and expressed as a locus of emotional action. In Chinese colloquial idioms, we encounter the red heart of love and youth; the hurt and broken heart; the erect and pure heart; the heart carried inside the mouth (i.e., the talkative heart); the cold, freezing heart of disappointed love; the heart-fire of congested anger, etc. These idiomatic expressions have parallels in German, e.g., the heart laughs inside my *Leib*, or the heart jumps inside my *Leib*; to pour one's heart, so it may not burst; grief that does not sprout oppresses the heart until it breaks.[8] We can find many expressions like these in other languages. But there is a small, yet decisive difference between the Chinese and German idioms of the heart. In Chinese the rhythmic action, e.g., the pounding, beating and jumping of the heart of joy and happiness is missing. If the heart jumps or palpitates (*xintiao, xinji*) it refers to something negative, refers to a heart in fear and danger (Ots 1990b). The cultural motto of quietness and harmony that constructs a still and empty heart has left its imprint on Chinese idiomatic language. Yet cultural restriction apparently does not keep the cathartic person from feeling how "fountainwater-like tears wash away all the misery which was buried deep inside the heart."

The social context of spontaneous *qigong*

The Great Proletarian Cultural Revolution (1966–78) inflicted serious suffering on many Chinese. In its later years, at a point of great confusion about who was fighting against whom and why, *leib*ly perceived depression became a dominant state for many people. In interviews conducted between 1978 and 1980, I was told repeatedly that it was advisable to "sit still," because every action might bring one trouble.

In 1978, in Beijing, a spontaneous mass movement referred to as the "Democratic Movement" came into existence. Its most famous site for testimony was the "Democratic Wall," located in the heart of Beijing. On it, individuals plastered wall posters decrying their fate or demanding political change. The people of China "stood up" and were on the move as Beijing witnessed a series of spontaneous mass demonstrations that demanded political and also cultural changes. This period of political protest was stopped when the new leadership under Deng Xiaoping realized that the movement was going too far and constituted a threat to the power elite. In the fall of 1979, the "Democratic Movement" was brought to a violent halt, some of its leading figures were persecuted, the "Democratic Wall" was outlawed and, in a sweeping night action, stripped of all its posters.

It is against this background of a restoration of communist rule and failed hopes for a major political change that the *qigong* movement came into existence in 1979/1980. Now that it was virtually impossible to voice one's political ideas, spontaneous-*qigong* served as another means of self-expression. The following years brought ever more severe phases of political restoration (1983, 1986, and 1989). Immediately after each of these rollbacks, the groups of people who practiced spontaneous-*qigong* in the parks diminished. But sometimes they grew again in size. When in November 1989, five months after the "Beijing Massacre," I revisited the Earth-Temple Park in Beijing which in the years before was a favorite location for hundreds of practitioners of spontaneous-*qigong*, it was hard to find even a small group (of about fifteen people) practicing crane-*qigong*. Now, however, when practitioners reached the point where they previously would have engaged in spontaneous movements, they remained still and absolutely silent. There was only one man who engaged in small, circling movements of the hips. He was totally in control and, as if afraid of being watched, he cautiously observed the surroundings. There was no trace of the emancipatory air of self-expression and joy of prior years. When I left the park, I stopped and took a few deep breaths of air that gave some temporary relief to my constricted chest. Then I shook myself and hurried away.

Mary Douglas (1982 [1970]: 74), commenting on the relation between society and motion, said "We tend to find trance-like states feared as dangerous where the social dimension is highly structured, but welcomed and even deliberately induced where this is not the case." The Chinese *qigong* experience decisively supports this claim. The leader of the above-mentioned Nanjing crane-*qigong* group had to cope with many attacks from the local office of public security as well as the local *qigong* organization. He was attacked because his group's form of *qigong* was the wildest, allowing even members of different sexes to touch each other. He personally believed he was attacked because he resisted giving his group any formal organizational structure, charged no fee for his services, and refused to join the local *qigong* organization. "Organization means rules and control," he told me. "Under organizational control we could not move as freely as we do now. After a short time, our group would cease to exist."

In order to bring the growing *qigong* movement under its control, beginning in 1990, sweeps have been directed towards the groups of *qigong* practitioners in the parks. Their leaders were forced to register and prove their conformity with the legitimized norms of *qigong* practice. *Qigong* gatherings of more than 1,000 adepts have been banned after up to 15,000 frenzied participants who writhed, screamed, laughed or cried as masters projected *qi* upon them took part in Beijing *qigong* rallies.

In December 1989, sponsored by the Ministry of Health and the Beijing Administration of TCM, the "World Scientific Association of Medical *Qigong*" was founded in Beijing, and an international conference was held on that occasion. A few hundred papers were delivered covering a wide range of different forms of *qigong* and their related healing effects. Spontaneous-*qigong*, which during the whole period of "*qigong* craze" had been the most popular one, was not even mentioned once.

Leib, spontaneity and millenarian movement

Forty years of unsuccessful attempts to construct a "new man of socialist morale" have curtailed many ancient traditions and ways of life without providing new directions. Within this vacuum, the practitioners of quiet and still forms of *qigong* formed a communitas of withdrawal and retreat. Crane-*qigong* was also composed as a quiet form. But then the unexpected happened: practitioners lost control and fell into ecstasy. Without spontaneous-*qigong*, the "*qigong* craze" would not have come into existence. Soon after, many practitioners of *qigong* expressed a longing for supraordinary intelligence, superpower and supranormal abilities (*teyi gongneng*). *Qigong* journals display, in a series of comic strips, such longings: old and feeble persons uproot trees, propel their wheelchairs, merely by projecting their *qi* (*fa qi*). These notions reveal some aspects of a millenarian movement in its early phase. In Chinese history, the majority of millenarian movements (whether we look at the Yellow Turbans of the second century AD, or at the Boxer Movement of the nineteenth century) were intimately connected with the experience of different forms of *leib*ly states, especially ecstasy and trance which seem to have been connected with feelings of invincibility. Williams's (1923) account of the Melanesian Vailala madness reveals phenomena which are very similar to those of spontaneous-*qigong*, and which were phrased by the natives as "head-he-go-around," "whirling motion," "whirlwind," "belly don't know," "dizzy," "ecstatic," and "crazy." Williams wrote:

The natives ... were taking a few steps in front of them, and would then stand, jabber and gesticulate, at the same time swaying the head from side to side; also bending the body from side to side from the hips, the legs appearing to be held firm. (Quoted in Worsley 1986: 75–6)

I submit that millenarian movements do not merely replace a deficient cultural mazeway or a broken cultural structure with a new one. The creation of a new orientation is anticipated by a phase of *leibly* spontaneity during the liminal phase. Persons are stripped of much cultural/social structure, and pushed back into themselves, into their *Leib*. Mary Douglas (1982 [1970]: 73) points out this central feature of millenarian movements:

[M]ost revivalist movements go, in an early phase, through what Durkheim called "effervescence" ... emotions run high, formalism of all kinds is denounced, the favored patterns of religious worship include trance or glossolalia, trembling, shaking or other expressions of incoherence and dissociation.

It seems that the experiencing of the expressive *Leib* generates new ideas in times of sociocultural decay. However, the "*qigong* craze" cannot be labeled as a millenarian movement yet: it still misses a cohesive ideology and a charismatic leadership in the Weberian sense.

One can thus analyze the *qigong* craze within the framework of communitas, liminality, structure and anti-structure (Turner, 1969). The *qigong* craze was spontaneous and immediate, and as such, opposed to the norm-governed institutions in place as social structure. The craze began during a period of sociocultural transition, its actors were liminal personae (threshold people). Eventually, it too became institutionalized; *Qigong* organizations sprang up which tried to lead its followers into some realm of state-approved cultural values. Turner defined communitas as speculative, generating imagery and philosophical ideas; the early communitas is frequently associated with mystical power, and there is something "magical" about it. He stressed its human interrelatedness. The *qigong* craze reveals an aspect of communitas which is interrelated as well as "individual": the lessening control of mind over body, and the rise of *leibly* spontaneity. Turner (1969: 138) wrote of ecstasy that it is "to stand outside the totality of structural positions one normally occupies in a social system." There is another side to Turner's coin, however. That is to stand outside one's mind, to let the *Leib* take the lead in an experience where, to repeat the words of one of the practitioners cited above, the "movements no longer follow the thought; instead, the thought and images develop out of the movements and are guided by them."

The bifold relation between the *Leib* and culture

In the contemporary Chinese context, the framework of public discourse left little if no positive space for emotions. Because emotions are understood to make the body move in *qigong*, the movements themselves became stigmatized. The cultural values of control, quietness, relaxation, and harmony turned against a body in e-motion. Mary Douglas (1982 [1970]: 65) writes:

the social body constrains the way the physical body is perceived ... As a result of this interaction, the body itself is a highly restricted medium of expression. The forms it adopts in movement and repose express social pressures in manifold ways.

I want to go further, and suggest that the structure of culture is homologous with the basic ways our *Leib* is perceived. It frames, diminishes, or even restricts our perception, awareness, and consciousness of the *Leib*. Under normal (cultural) circumstances, the *Leib* is thus hindered in its self-expression. However, even in special situations (e.g. the cathartic effect of *qigong* in our case) when the *Leib* is set free from cultural constraints, it becomes freed in culturally recognizable ways. Whether the *Leib* expresses itself in a cathartic fashion or not depends on the individual situation and the sociocultural context. Culture organizes our means of public expression of *leib*ly renderings. In China, it forces people to communicate within a discursive style that does not threaten culture and culture's advocate within us, the mind. In public discourse, Chinese people do not speak of their emotions, but employ cultural metaphors such as *qi* which were once grounded in *leib*ly phenomena. With the passage of time, the metaphors have lost their *leib*ly content.

What are the implications of this for anthropological research? Isn't it that we anthropologists seem to be satisfied and happy if, in fieldwork, our findings coincide with what we have learned earlier about that specific culture? Usually we are satisfied with the first set of answers given to us by our informants: the data suit our expectations. Personal perception, thought, and behavior seem to merge with culture in a single dimension: the culturally inscribed and constructed body. And off we go! But such an approach would only entitle us to say that people employ culturally and socially valid styles of public discourse. It can tell us something about the body social, but little about the body self, the *Leib*. To paraphrase Merleau-Ponty (1964), we should remain aware that "primacy of perception" necessarily implies "privacy of perception." Fanny Cheung (1982) has repeatedly raised her voice to point out the more "inside," private levels of perception and discourse that can so easily evade the Western anthropologists' attention. Research on culture and *Leib* thus necessitates the intimate relation between researcher and informant, and repeated interviews – or better, informal talks – with the same persons.

My research in *qigong*-healing suggested several ways in which culture and mind oppose the *Leib*. But I remain unsatisfied with my findings with respect to the *Leib*'s part in constructing culture, how *leib*ly phenomena become systems of meaning. Our enculturation and socialization have alienated us from our own *leib*ly feelings. Following Csordas (1990: 9), I believe it is now "necessary to allow us to study the embodied process from beginning to end instead of in reverse" because "phenomenology is a descriptive science of existential beginnings, not of already constituted cultural products." I argue that it is time to reconsider our epistemological tools: the *Leib* cannot be thought of, it must first be experienced. This calls for an approach in *Leib* research where one goes beyond participant observation – "experiencing participation" would be more to the point. For example, I entered the group as a follower of *qigong* rather than as an anthropologist. After a short time, I tranced myself. I am convinced that experiences like these helped me to understand better the *leib*ly basis of frequently used Chinese terms. This includes the well-known symptom and feeling of *xiongmen* (thoracic depression (Ots 1990b)) which practitioners reported suffering prior to their engaging in *qigong* practice. Thoracic depression (in connection with fright) was also the most often-perceived transitory *leib*ly feeling reported when hidden emotions erupted, yielding subsequently to the feeling of opening up and of expansion. In Chinese, being happy can be expressed by the term *kai xin* (open one's heart). This e-motional expansion was enacted and demonstrated by the movements. Expanding emotions made the practitioners leave their prior position. They floated around the practice ground, movements became light and rhythmic, and very often they would raise and swing their arms and start to shout and laugh.

My observations of this existential connection between depression and expansion are admittedly only minute details in our rediscovery of the *Leib*. What is needed is personal experience which makes possible a "thick description" of *leib*ly processes.

ACKNOWLEDGEMENTS

I want to thank the practitioners and instructors of the Nanjing crane-*qigong* group for giving me the opportunity to share their experience and for allowing me to bother them with my repeated questions. I greatly appreciate the many helpful comments and criticisms I received from Thomas Csordas, Eva Fulde, Susan DiGiacomo, Gudula Linck and Boris Voyer. Although their helpful suggestions have shaped this article, I am fully responsible for its content. I am grateful to the Deutscher Akademischer Austauschdienst who gave financial support to my studies in Nanjing in 1984–5. Last but not least, I want to thank Christa Carjaval for polishing my badly knitted English prose.

NOTES

1 Remarks made at conference on "Medical Anthropology: North American and European perspectives," Hamburg, Germany, December 1988.

2 Schmitz's work on the *Leib* is concentrated in his ten volume *System der Philosophie*, published between 1964 and 1980.
3 Traditional Chinese medicine knows seven emotions. These are: anger, joy, worry, sadness and grief, fright and fear.
4 The term *qigong* is rather recent (twentieth century). Older terms are *daoyin* or *yangsheng*.
5 David Holley in the *Los Angeles Times* 16 October 1990: H8.
6 I follow Thomas J. Scheff's (1979: 14) definition of catharsis as a therapeutic process that discharges repressed emotions.
7 *dong qing* = the Chinese words are movement and feeling.
8 In Wander's *Deutsches Sprichwörter-Lexikon* (German Idiomatic Dictionary) of 1870 we find 573 heart-related idioms.

REFERENCES

Cheung, Fanny (1982) Psychological Symptoms among Chinese in Urban Hong Kong. *Social Science and Medicine* 16: 1339–44.
Csordas, Thomas (1990) Embodiment as a Paradigm for Anthropology. *Ethos* 18(1): 5–47.
Douglas, Mary (1982) [1970] *Natural Symbols*. New York: Pantheon.
He Zongyu (1985) Mawangdui guyi yu zhongguo chuantong de kangfu yixue (The Recordings of Mawangdui and China's traditional Rehabilitation Medicine). *Qigong yu kexue* 12: 9–11.
Leder, Drew (1984) Medicine and Paradigms of Embodiment. *The Journal of Medicine and Philosophy* 9(1): 29–44.
Li, Chuncai (1983) Shilun qiong zhong de dongxiang (Thesis on the Movements in qigong). *Qigong zazhi* 4: 169–72.
Linck, Gudula (1991) Das Zeichen für Herz – Zwischen Vernunft und Gefühl. Überlegungen zu einem Chinesischen Begriff. Unpublished manuscript.
Lü Bingkui (1980) Zhuhe "qigong zazhi" chuangkan (Congratulations on behalf of the publishing of "qigong zazhi"). *Qigong zazhi* 1: 1–3.
Ma Chunyi, Wang Xiaobin, and He Jiwen (1988) Zifagong – meng de bian zhong (Spontaneous qigong – a kind of dream transformation). *Qigong yu kexue* 5: 25.
Merleau-Ponty, Maurice (1964) *The Primacy of Perception*. London: Routledge and Kegan Paul.
Ots, Thomas (1990a) *Medizin und Heilung in China – Annäherungen an die traditionelle chinesische Medizin*. Berlin: Reimer (2nd edition).
  (1990b) The Angry Liver, the Anxious Heart, and the Melancholy Spleen: The Phenomenology of Perceptions in Chinese Culture. *Culture, Medicine and Psychiatry* 14(1): 21–58.
Qi, Hao (1988) Qigong re; zhongguo qigong zouxiang shijie (Qigong-Fever: China's qigong meets the world). Beijing: Library of the Beijing College of Traditional Chinese Medicine.
Robinet, Isabelle (1989) Visualization and Ecstatic Flight in Shangqing Taoism. In Livia Kohm, ed., *Taoist Meditation and Longevity Techniques*. Ann Arbor: Center for Chinese Studies, The University of Michigan – Michigan Monographs in Chinese Studies, Vol. 61, pp. 159–92.
Rosaldo, Michelle Z. (1984) Toward an Anthropology of Self and Feeling. In

Richard A. Shweder and Robert A. LeVine, eds., *Culture Theory – Essays on Mind, Self, and Emotion.* Cambridge: Cambridge University Press, pp. 137–57.

Scheff, Thomas J. (1979) *Catharsis in Healing, Ritual and Drama.* Berkeley: The University of California Press.

Scheler, Max (1962) [1928]. *Die Stellung des Menschen im Kosmos.* Bern und München: Francke.

Schmitz, Hermann (1989) *Leib und Gefühl – Materialien zu einer philosophischen Therapeutik.* Paderborn: Jungfermann.

Schwartz, Benjamin I. (1985) *The World of Thought in Ancient China.* Cambridge, MA: Harvard University Press.

Turner, Victor (1969) *The Ritual Process – Structure and Anti-Structure.* Ithaca, NY: Cornell University Press.

Williams, F. E. (1923) *The Vailala Madness and the Destruction of Native Ceremonies in the Gulf Division.* Papuan Anthropology Reports, No. 4, Port Moresby.

Wang Guoliang (1985) Hexiangzhuang zhong zhong (All kinds of Crane-qigong). *Zhonghua qigong* 3: 44–5.

Worsley, Peter (1986) *The Trumpet Shall Sound: A Study of Cargo Cults in Melanesia.* New York: Schoken Books.

Part III

Self, sensibility, and emotion

6   Embodied metaphors: nerves as lived experience

Setha M. Low

As medical anthropologists we struggle to understand sickness with our disciplinary tools, those of the analysis of the social, cultural and political relations that produce sickness. But in this struggle we have neglected the biological body of process, function and structure and the physical body presented to us and inhabited by us in everyday life. This neglect has occurred partly because of the limitations of disciplinary techniques and epistemological traditions, partly because the concern has been to bring the social, political, economic and cosmological significance of sickness to the attention of the medical world, and partly because the biological and physical aspects of the sensing body simply can not be as easily seen or talked about by the sociocultural researcher.

One example of how lived experience is excluded from our analyses can be found in my own work on *nervios* (nerves), a form of suffering found in many Western cultures. The inchoate quality of the experience of *nervios* is not adequately captured in the analysis even though the sufferers use an elaborate cultural vocabulary of signs and symptoms. My analyses of *nervios* focused on this vocabulary, the idioms of distress and their meanings used by sufferers, in Costa Rica (Low 1981; Bartlett and Low 1980, 1988) and Guatemala (Low 1988). When reviewing this work a dilemma emerges of how to reduce the dualism of mind and body, sensation and sense, biology and culture that is imposed by the analysis.

Within medical anthropology recent critical interpretative-hermeneutic approaches provide new ways to think and write about the body as it is experienced (Scheper-Hughes and Lock 1986), whether it is through figurative language (Turner 1984, Lock and Dunk 1987; see also Johnston et al. 1985), narrativity (Good 1990), or accounts of torture and terror (Jenkins 1991; Scheper-Hughes 1989; Taussig 1987). Alternatively, Csordas (1990), Frank (1986), and Gordon (1990) resolve the analytic dilemma through the paradigm of embodiment grounded in perception and practice "that leads to the collapse of the conventional distinction between subject and object" (Csordas 1990: 40). These recent innovations are particularly instructive in that they suggest a way in which a reconceptualization of nerves-suffering

139

that focuses more on the lived experience and less on the vocabulary of signs and symptoms might be possible.

Nervios/nerves sufferers talk about what they are experiencing in a variety of ways: they say that they feel as if they are trembling or shaking all over, that they are jittery or that their skin seems to jump up and down. They complain that they feel unusual sensations such as hot or cold enter their bodies, or that they feel tired, as if their energy was drained away and their bodies were deflated. Some individuals say that they lose consciousness, faint or feel outside of themselves. In almost all cultural contexts, sufferers express the *nervios*/nerves experience as feeling "out of control" as if their bodies were moving, contracting or disappearing, and were no longer theirs. These physical/emotional sensations, which are discussed in greater detail in the following case studies, have been analyzed as "symptoms" of nerves.

Sufferers, on the other hand, refer to these sensations as signs or indicators of nerves, using these descriptive phrases to tell friends, physicians or anthropologists what they are experiencing. Sufferers sometimes identify one sensation as nerves in one context and as some other entity in another; in most cases the entire complex of sensations is identified as nerves. In my own research it has often been unclear whether the person meant that the sensation was nerves or that the sensation indicated that they had a case of nerves, and this indeterminacy of the relation of the sufferer to nerves remains obscure. The language of symptoms and diagnosis removes the sensations from the physical and biological body of the sufferer, obscuring the person/body/experience relationship. Drawing upon the recent literature on embodiment, these sensations can be reconceptualized as the physical/emotional embodiment of nerves. This reconceptualization places the sensations back in the body of the sufferer and in most cases demedicalizes them.

The research that describes the experience of *nervios*/nerves sufferers also reports a recurring set of sensations expressed in metaphoric language. *Nervios*/nerves sufferers shake like a leaf, feel "on pins and needles," they lose themselves or lose parts of themselves, feel numb as if they were dead or feel as if they can not go into the street. Often there are sensations that are perceived as not part of the body such as "my arm felt as if it did not belong to me," or sufferers say that they are outside, on the surface or below their bodies. Often the language of the sufferers evokes other bodily metaphors such as those of heat, expansion, or breakdown, often expressed in hydraulic language. For example, Lock describes *nevra*, nerves of Greek women in Montreal, as a feeling of "bursting out," "breaking out," or "boiling over," "an experience, therefore, of crossing the 'natural' boundaries between the inside and out" (1990: 238). These metaphors have rarely been discussed in terms of their experiential impact and raise a number of important questions: What does it mean to say that a person feels like "shaking all over," "bursting out," or "boiling over"? Do these metaphors provide clues to bodily experience? What is the locus of these metaphors? Are illness metaphors specific cultural means of communicating? If so, how does each culture frame the metaphorical expression and experience of nerves? Ultimately, are these metaphors essential to our understanding of cultural variation in illness embodiment?

In an early article I argued that *nervios* in urban Costa Rica was a metaphoric statement to be understood in contrast to "being tranquil," a dominant cultural goal (Low 1981). This metaphoric analysis seems correct for Costa Rica. However, in a later article comparing nerves in different cultures, I noted that although the reported bodily sensations were similar in a variety of cultures, folk explanations of etiology and cultural meaning varied considerably (Low 1985). I concluded that *nervios* was a culturally interpreted symptom – as opposed to a culture-bound syndrome – suggesting that the metaphoric power of nerves is more culture specific than the reporting of the bodily experience. The metaphoric language of nerves embodiment thus refers to local cultural usage and meaning, even though the embodiment may have cross-cultural threads of common lived experience.

There also has been extensive scholarship documenting the conditions that give rise to nerves with explanations ranging from those related to the breakdown of family and social networks, loss of loved ones and concern for the well-being of close friends and family members, to political analyses of gender, ethnicity and class inequality as well as of torture, disappearance and terror (Barlett and Low 1980; Low 1981, 1985; Davis and Low 1989; Davis 1988; Van Schaik 1989; Lock and Dunk 1987; Lock 1990; Jenkins 1988a, 1988b, 1991; Davis and Guarnaccia 1989; Scheper-Hughes 1989). Many of these works summarize the "meaning of nerves" in terms of it being a response to or a communication of distress presented in a culturally appropriate manner. An extreme example is Jenkins's (1991) recent article that describes *nervios* as an indigenously defined condition referred to in embodied accounts of the political ethos of *la situación*, the intolerable conditions of violence and poverty in El Salvador. Other studies such as Van Schaik's (1989), Scheper-Hughes' (1989) and Lock's (1990) suggest that nerves is a form of resistance, not just a response to or communication about sociopolitical and familial distress. Lock (1990), in the context of Greek women immigrants to Montreal, suggests that *nevra* (nerves) is a form of resistance to cultural expectations and explicitly enacted to "produce small victories for change" (1990: 250), yet experienced as "real" pain. Finkler (1989) goes so far as to argue that nerves is the embodiment of adverse existential conditions in the same way that a cold is the embodiment of a pathogen. She states that "It is my contention that nerves is not simply an idiom of distress, a metaphor for distress . . . Nerves signifies the *embodiment* of generalized adversity and recreates in the internal world of the body the perceived contradictions and disorder of the external world" (Finkler 1989: 174).

While I basically agree with Finkler that nerves is an embodiment of distress, I would argue that it is more complicated than equating nerves with distress. The embodiment is of nerves, and nerves is constructed by local discourse and institutions, then expressed and acted upon as a *metaphor* of social, psychological, political or economic distress. The relationship between nerves and the embodied distress, therefore, is culturally mediated, both in terms of what forms of distress cause suffering and in terms of its metaphorical expression.

Our somewhat limited focus on the cultural content of illness to the exclusion of its values and contextual reality has become part of a new politics of writing about nerves. Lock (1990) warns us in her article on the Greek ethnic identity of *nevra* that medical anthropology has

come dangerously close at times to giving fuel to a racist, "quaint ethnic" attitude, thinly disguised in many programs for multiculturalism. Rather than simply focusing on "nerves," perhaps we need to emphasize the losses and unresolvable contradictions which immigrants face, their own insights and efforts to deal with the situation, their continued exploitation as a work force, and *above all*, the unexamined values embedded in the "host" country. (Lock 1990: 251)

These values are embedded in the metaphors of nerves that are distinct to each cultural context, and these metaphors are embodied in acts of resistance as well as acts of compliance.

Combining the insights of the paradigm of embodiment – the physical/ emotional experience of nerves being the embodiment of distress – with the notion that metaphor conveys lived experience in a culturally meaningful way, repositions nerves as an embodied metaphor. In this reconceptualization, nerves embodies the lived experience of daily life as a metaphor of physical, social, political, and economic distress that has specific meanings within a variety of cultural contexts. In the following sections of this paper I elaborate the concept of embodied metaphor and then present seven case summaries of nerves research. In these cases the focus on symptoms and disembodied illness persist; however, in each I try to extract the embodied metaphor. In the final section, I attempt a preliminary comparison of the embodiment of nerves. This comparison suggests that there are some key similarities in the experience of nerves that provoke new questions about the universality of the embodiment of distress.

Embodied metaphors

"In sickness we confront the inchoate. Bodily suffering distorts the land-scape of thought, rendering our previous construction incoherent and incomplete" (Kirmayer 1992: 329). Metaphor, according to Fernandez (1986), is the primary way that individuals and cultures make sense of the world. Metaphor is a way to define the undefined and nascent identity of a person or group. Metaphor allows one to move from the abstract and inchoate of lived experience to the concrete and easily graspable. Metaphor is also strategic; it is a plan for action and performance. In terms of this discussion, metaphor allows for the communication of otherwise senseless and unspeakable suffering. Csordas states that the "performance [in ritual] of metaphors effects a qualitative transformation in participants. This is either a 'movement' from one state to another, or 'movement' from form-lessness or lack of identity to definiteness and specificity" (Csordas 1987: 459). Further, metaphors are creative and infinitely generative in their allusions and meanings. So metaphors can provide flexible, creative, and strategic language for the expression of suffering.

Cognitive scientists concerned with the development of thought argue that metaphor is the earliest vehicle for ordering bodily sensation and action (Johnson 1987). It is Lakoff's (1987) contention that rather than thought symbols deriving meaning only from their correspondence with things, reason is made possible by the body as it experiences the world, and grows out of the nature of the organism. Thought results from embodied experi-ence and is creative using metaphor, metonymy and mental imagery based on bodily experience. Thus, metaphor is grounded in the body and emerges from it, producing categories of thought and experience. Quinn (1987) has criticized Lakoff and Johnson for arguing that metaphors are solely bodily based and provides evidence that our metaphors for thought are culturally produced. All bodily sensations and basic movement are affected by culture through acquired habits and somatic tacts (Mauss 1950). Agreeing with both positions, the evidence suggests that metaphors of thought can be generated by the experience of the body as well as culture. Thus body metaphors provide a possible solution for the expression of lived experience that can communicate bodily sensation, as well as social, cultural and political meaning (Turner 1984; Foucault 1979).

Writing on the embodiment of metaphor, Kirmayer (1992: 335) argues that metaphor "offers a remedy for the drought of passion in semantics ... Metaphor theory shows how our purest, most abstract ideas are products of bodily action and emotion." Kirmayer (1988a) also sees the embodied metaphor as the solution to the semantic debate in medical anthropology in that metaphor maintains its grounding in the body, while at the same time providing the ability to extend into the social world. He warns us, however, that "the political control of expressive metaphors shapes our bodily experience" (Kirmayer 1988b: 29) as well, and that "This political process is at work in the patient's effort to make sense of suffering and the physician's interpretive acts" (ibid.). Certainly medicalization of social distress has become a political process based on creating medical metaphors out of everyday life.

Medical anthropologists such as Lock (Lock and Dunk 1987; Scheper-Hughes and Lock 1987; Lock 1988), Scheper-Hughes (Scheper-Hughes and Lock, 1986, 1987; Scheper-Hughes 1989, 1992), Martin (1987) and Pandolfi (1990) also have documented the power of the sociopolitical use of metaphor on the body, particularly for women and the poor. This sociopolitical manifestation of the embodied metaphor is dramatically portrayed in *nervios*/nerves (Davis and Low 1989) as well as in other metaphors of bodily affliction. Scheper-Hughes (1989, 1992) captures the subversive power of the embodied metaphor when she writes about the "somatic culture" of Brazil and the tendency of people to communicate through the body. Disturbances of the body, such as *nervos*, nerves, in Brazil are characterized in much the same way as *nervios*/nerves sufferers in other parts of North and South America, but Scheper-Hughes argues that these "nervous attacks" have political meaning:

[F]renzied nervousness associated with trembling, fainting and seizures, and paralysis of limbs, [are] symptoms that disrespect and that breech mind and body, the individual and social bodies. These nervous attacks appear to be in part coded metaphors through which the workers express their politically dangerous and therefore unacceptable condition of chronic hunger, and in part acts of defiance and dissent, graphically registering the afflicted one's absolute refusal to endure what is, in fact, unendurable and the protest against the demand to work, meaning always in this context, the availability for physical exploitation and abuse at the foot of the sugarcane. (1989: 7–8)

Scheper-Hughes's (1992) critique of the lost lives and disappeared bodies of northeast Brazil comes closest to a model of embodiment that includes the physicality of the body, its cravings, sensations and pain, yet at the same time retains the subversion and resistance of its metaphorical message.

Nervios and nerves

The phenomena of nerves, *nervios*, *ataque de nervios*, *nevra*, and *nervos* have been well documented in the literature in diverse Western cultures and subcultures (Low 1981; Barlett and Low 1980; Low 1985; Davis and Low 1989; Hill and Cottrell 1986; Davis 1988; Cayleff 1988; Van Schaik 1989; Dunk 1989; Lock and Dunk 1987; Davis 1989; Lock 1990; Jenkins 1988a,

1988b; Davis and Guarnaccia 1989; Guarnaccia, Rubio-Stipec, and Canino 1989; Guarnaccia, Good and Kleinman 1990; Dresp 1985; Scheper-Hughes 1989, 1992). Because *nervios* is so widespread, with diffuse symptoms and diverse etiologies, it has been the subject of a series of theoretical explorations that range from psychosocial to sociopolitical explanations of its frequent occurrence and cultural meaning. The emphasis of the analysis presented here, however, is on the experience of *nervios* among Costa Ricans and Guatemalans, *nervios* or *ataques de nervios* among Puerto Ricans, and nerves among individuals in Newfoundland and Eastern Kentucky. These embodied metaphors of suffering retain the integration of mind/body experience and express the physical as well as the social and political basis of *nervios*/nerves.

Costa Rica

Nervios has been studied in three regions of Costa Rica: in two rural communities located in the western Puriscal Canton of Costa Rica (Barlett and Low 1980), in the hospital clinics of San José, the capital city of the country (Low 1981), and in a rural community on the English-speaking Caribbean coast (Hill and Cottrell 1986).

Rural Puriscal

The rural study was based on open-ended interviews and informal discussion with all households in two agricultural communities located about a two-hour bus ride from San José. Information on *nervios* was gathered from both the husband and wife in all seventy-five households in the first community. A total of 107 persons reported having experienced *nervios*, comprising 25 percent of the total population of 430 persons. In the second community a research assistant surveyed all thirty-eight households. She found 68 cases of *nervios* out of 199 persons, slightly over one-third of the community.

The label of *nervios* was given to an extremely wide range of sensations including fear, trembling, crying, upset stomach, pounding heart, dizziness, feeling strangely hot or cold, insomnia, and loss of appetite. One old woman reported "I can be as tranquil as can be, and all of a sudden, from nothing, I get *nervios*. My body shakes all over and my heart pounds, and my skin prickles. It's been like this all my life" (Barlett and Low 1980: 526). Nerves also includes conditions such as headache, "brainache," ulcer, allergy, backaches, high blood pressure, and asthma. One girl recounted how she had had *nervios* since childhood. "One day, after bathing, she came into the kitchen and the kitchen table 'was spinning.' She went to bed, but she felt disoriented and 'didn't know who she was.' When she got up, she hit her

head hard against the wall ... and exclaimed, 'these *nervios* are terrible!' "
(Barlett and Low 1980: 527).

Some *nervios* is said to be caused by a short-lived event such as the death
or injury of a family member, or a trivial event such as pricking a finger on a
needle. Other interviewees suggest that a certain event may bring on the
nervios but attribute the person's susceptibility to more basic causes such as
weakness or heredity. *Nervios* is commonly said to come through the
generations (*por generaciones*) when the complaint is chronic. In the rural
context the cause of *nervios* is perceived as biological or genetic, but the
incident is triggered by social or emotional events. It is explained as a
disease of the nerve-endings or of the brain which causes a person to feel or
act inappropriately.

Nervios in the rural Costa Rican context is found more commonly among
women, among married people, among parents, and among those aged 19
through 60. It occurs more frequently during periods of family formation
and contraction. The opposite of "having *nervios*" is "being tranquil"
which is a cultural goal expressed by all Costa Ricans. Barlett and Low state
that there are three important points about *nervios* in these rural communities.

First, the sufferer is not responsible for the symptoms of the complaint ... Second,
by labelling distress as *nervios*, those around the sufferer are also absolved of
responsibility ... Third, the fact that *nervios* is viewed as a biological and sometimes
hereditary trait, removes the sufferer from any responsibility to seek further causes
of the distress and obtain a remedy. If *nervios* symptoms persist or are severe,
sufferers often seek medical care, but there is no emphasis on determining or
eliminating any stressful conditions linked to the condition. (Barlett and Low 1980:
552–3).

In rural Costa Rica, then, *nervios* is embodied distress related to the pain
and suffering of creating a family, that is, getting married and having
children and losing family members through death or children moving
away. *Nervios* bodily manifestations absolve the sufferers from responsibility for their bodily transgressions and social acts. It derives metaphoric
meaning from its opposition to living a tranquil life, and is experienced as a
wide range of physical and emotional sensations expressed in graphic and
vivid bodily terms.

Urban Costa Rica

The research data for the urban Costa Rican study were collected in
outpatient clinics of four hospitals within the two principal Costa Rican
health-care delivery systems. These clinics serve a majority of urban
patients living in San José and the surrounding suburbs and towns, but also

are visited by rural residents who prefer the health institutions of the city. The data on *nervios* draw upon various methodological sources which include 457 cases of observed doctor–patient interaction in medical consultations, 117 interviews with patients before and after their medical consultations, 12 team-conducted family interviews in the patients' homes, and twenty months of participant observation in three neighborhoods of San José (Low 1981). The final clinic sample included 305 patients from general medicine clinics and 152 from psychiatric and psychosomatic clinics. Seventy percent of the patients in general medicine and 63 percent in the psychiatric clinics were female, with a total patient mean age of 33.5 years.

Of the 457 patients in the sample, 122 complained of *nervios*; it was the most commonly presented complaint in the psychiatric clinics (50 percent of all patients observed) and the second most common complaint in general medicine (15 percent of all patients observed). These patients exhibited a pattern of reported sensations associated with their *nervios*: headache, insomnia, lack of appetite, depression, fears, anger or bad character, trembling, disorientation, fatigue, itching, altered perceptions, profuse sweating, lifelessness, vomiting, and hot sensation. The person commonly feels "out of control," or separated physically and emotionally from her/his body. They complain that the sensations are not part of their normal behavior, but are experienced as undesirable body responses over which they have no control.

The description of people feeling "out of control" is reiterated in the before and after interviews. A young man from the rural highlands heard a noise in his head, was constantly dizzy, and had bouts of sweating, fear, sense of pressure, and neck pain. "I lost control," the patient told the interviewer, "and blacked out twice. I am concerned my *nervios* will cause this again." A middle-aged woman from a southern valley of Costa Rica, with nine living children, told us about her headache, dizziness, crying, and temporary blindness. A week before she had an attack of *nervios, a derrame de cerebro*, during which she was unconscious for two hours. She thought it might be family worries and hoped that the doctor would be able to help her as she felt as if she could no longer control herself. A single man from a working-class neighborhood employed in a laboratory department of a large hospital said his *nervios* was characterized by anxiety, desperation, and being disoriented. He was lonely and did not feel that people liked him at his job. He was worried that his disorientation would make it difficult for him to return to work (Low 1981: 34).

Nervios patients differed from the general medicine sample in that they experienced more family disruption. They are more often single, separated, or widowed, and more frequently mentioned family or spouse abandonment, death, or abuse in relation to their *nervios*. *Nervios* was repeatedly attributed to family interactional discord, disruption of family structure, and past family disturbance. Doctors also attribute *nervios* to family factors, but suggest that economic problems, spouse inattention and abandonment, lack of friendships outside the family, overdependency, boredom, and sexual problems are other causes. Doctors responded positively to patients who presented *nervios* and often asked them about their problems and concerns.

Nervios in urban Costa Rica is an embodied metaphor of the self and the self's relation to the social systems expressed through a disturbance or "discontinuity" of the body perception. It embodies the breakdown of the individual's relations in the social system and his/her role as a cultural participant; its embodiment elicits help for individuals and families experiencing distress and attempting to re-establish their normal sociocultural state. For many sufferers the disturbed body sensations of *nervios* are the embodiment of broken family ties and relations which in the urban context are the basis of identity, thus, ultimately *nervios* embodies the experience of a loss of self. Sufferers express this interpretation rather directly by saying "I lost myself" or "I could not tell where I was, I was so dizzy and confused." At least in this example, the embodiment brings together the experience of social, cultural, and existential loss in one metaphoric statement.

Caribbean Costa Rica

The data on *nervios* were collected in Caribe, a rural community of about 1,500 located in Limon Province on the Atlantic coast of Costa Rica (Hill and Cottrell 1986). Structured interviews focusing on mental disorder and *nervios* were conducted with a 40 percent random sample of West Indian (English speaking) households. In addition, key informants were interviewed about their perceptions of changes in the health and religious beliefs in their community. Systematic elicitation from informants using unit linguistic contexts or "frames" produced a categorization of mental disorders from a sample of 13 persons (Hill and Cottrell 1986).

Based on the linguistic elicitation technique, a person with *nervios* was characterized as having trembling hands or body, nervousness, jumpiness, as disliking noise, worrying a lot, crying a lot, or being so nervous they can not work (ibid.: 5). One respondent described her *nervios* as follows: "If I'm thinking, worrying about something, my hands and feet sweat and get cold. Then I worry about that. My body feels like trembling. Through tension, my head gets dizzy, wants to hurt. When *nervios* used to bother me, I felt uneasy, keep moving, start and can't finish, not contented, need to talk to someone" (ibid.: 7).

Over 35 percent of the interviewed sample suffered from *nervios*, with women having a higher incidence than men. Treatment forms varied according to perceived severity: 50 percent received medical treatment from a physician, 7 percent went to the hospital, 14 percent rested or otherwise changed their routine, 11 percent used a folk remedy, and 36 percent did nothing. Doctors, when visited, gave tranquilizers to 35 percent of the interviewees and vitamins to 7 percent (ibid.: 7). *Nervios* was thought to be caused by something unbalancing the body such as too much worry, alcohol or strong emotion, or inadequate diet, and having problems that cause one to be upset. While *nervios* is used to communicate distress to family members and friends, it seems to have more stigma attached to it and does not elicit the support and concern reported for other parts of Costa Rica (ibid. 1986).

The embodiment of *nervios* in Caribe retains many of the elements of the previous Costa Rican studies, including the trembling and dizziness. However, the metaphor in this English-speaking Caribbean population is one of being out of balance rather than being out of control. Further, the embodiment of being out of balance, whatever the cause, does not give the sufferer the absolution and support that nerves usually elicits. It is as if nerves may still be an embodiment of distress, but it is an embodiment that is perceived as unacceptable in the Caribbean coast context. Unfortunately, the research published on Caribe *nervios* does not detail the underlying distress except in the most general terms. Hill, however, reports rapid social change and increased community conflict and breakdown over issues of local political control as well as strained relationships between the mestizo Spanish-speaking migrants to the Caribbean coast and the black English-speaking residents (personal communication).

Guatemala

The Guatemalan study of *nervios* is part of a larger community survey on the growth and cognitive development of urban children in El Progreso, a resettled *colonia de reconstrucción* located about two kilometers from Guatemala City (Johnston et al. 1985; Johnston and Low, forthcoming). The *nervios* data were collected during one complete cycle of the interview process and included the entire sample of families measured in one year (June 1982 through June 1983). The questions were answered by the mother of the child who was being monitored, but in some cases by the older sibling or father if present during the interview (Low 1988).

Trained Guatemalan research assistants interviewed 362 families, of which 322 were utilized, representing symptom reports on 308 females and 207 males for a total of 515 individual cases. Three hundred and twenty-seven of these individuals or 63 percent reported having *nervios*, although the distribution varied by sex. Of the 308 females, 269 or 87 percent reported that they suffered from *nervios*, compared to 53 out of 207 males or 26 percent.

The sensations associated with the incidence of *nervios* included headache, despair or desperation (*desesperación*), trembling, face pain, teeth pain, bad humor (*mal humor*), anger, dizziness (*mareas*), stomach ache, twitching or jerking of the body (from the verb *brincar*), high pressure, and low pressure. The attributed causes of *nervios* were *cólera* (anger, fury or rage), *pẽna* (punishment, affliction, grief, or sorrow), birth control pills, work problems, death of family members, *preoccupaciones* (worries), birth of a child, *susto* (fear), problems, and pregnancy. Thirty-four individuals reported that they didn't know the cause of their *nervios*.

The Guatemalan interview schedule separated the questions of the initial cause of *nervios* from a question on when it occurs. In response, 40 percent of the women and 36 percent of the men stated that anger triggers their *nervios*; a common response was "I get them (*nervios*) when I am angry."

The high reported incidence of *nervios* and the predominance of anger and emotion as its etiological explanation suggest that *nervios* is embodied distress, or at the very least a response to a stressful social, political, and economic situation. El Progreso is a recently resettled community; most of the residents either lost their homes during the 1976 earthquake or have migrated from the countryside to work in the city. Political unrest, terrorism, and violence as well as a severe economic recession further constrain individuals' lives.

Studies of terrorism and violence have shown that *nervios* can be a common symptom expressing the rage and fear of potential victims and their families (Sluka 1989; Taussig 1987; Jenkins 1991). Residents, in the same conversations that they discuss their *nervios*, will tell stories of the loss of loved ones from violence in the central market and of the feared police who cruise by the community. A resident said that she thinks that there are more *nervios* problems in the city.

We are surrounded by cars. Suddenly you are in the street and you hear gun shots; all this makes you nervous. You are in danger and have to be alert. In danger not only from the bullets, but even when you cross the street. Small town life is more tranquil and therefore healthier. (Low 1989: 38)

But *nervios* is more commonly discussed in terms of the anger and sorrow that is felt with the loss of a child, or the irritation and exhaustion experienced from children making too much noise. Residents commonly refer to their anger, rage, or fury as causing their illness and if not anger, then pain and suffering. Another resident said that *nervios* comes because you have a problem and get tense. You get *nervios* and you get a headache:

For instance, one is busy and a child gets hurt. You worry and run to find out what happened. You torment yourself and hit the children. Later you come down with a headache from *nervios* because you got tense. Many times it is because the person is a very angry person. (Low 1989: 37)

In the Guatemalan case, *nervios* embodies distress at a personal, familial, social, and political level. The incredibly high incidence of this embodiment, 63 percent of the community and 87 percent of the women suggests that along with the cases of families in war-torn Ireland (Sluka 1989), Salvadoran refugees in the northeastern United States (Jenkins 1991), rural residents of northeastern Brazil (Scheper-Hughes 1989, 1992) and other victims of violence, discrimination, and political oppression the residents of El Progreso suffer both from their own personal sorrow and from the sociopolitical conditions of their everyday life. Yet the metaphor of *nervios* in Guatemala is expressed in terms of emotion, particularly anger and suffering, that retains elements of more traditional indigenous illnesses such as *susto*. These indigenous roots could be interpreted as passive resistance. More generally though, *nervios* embodiment does not generate acts of political resistance or defiance, but is passively experienced through the body.

Puerto Ricans in New York City

Harwood (1981) in his book on *Ethnicity and Medical Care* compiled all available sources on Puerto Rican health behavior, beliefs, and practices into one readable chapter. From this source it is possible to get some sense of the role of *nervios* in the urban Puerto Rican community, how it is presented and how it might be interpreted. More recently Peter Guarnaccia and colleagues (Guarnaccia, Good, and Kleinman 1990; Guarnaccia, Rubio-Stipec, and Canino 1989; Guarnaccia, DeLaCancela, and Carrillo 1989); Koss-Chioino (1989); and Christine Dresp (1985) have published ethnographic and epidemiological studies of *nervios* and *ataque de nervios* among various Puerto Rican populations in the Eastern United States and on the island of Puerto Rico. This summary draws upon the findings of these sources to present an overview of Puerto Rican *nervios* and the more characteristic *ataque de nervios*.

Nervios was the third most common symptom presented by adults in a general medicine clinic in New York City, representing 6 percent of the 96 persons interviewed. The prevalence of *nervios*, however, is probably underestimated by these statistics in that *nervios* is a condition likely to be treated at home with herbal and over-the-counter preparations or by culturally acceptable spiritual healers. Nerves is most commonly referred to by Puerto Ricans as an *ataque de nervios* (attack of *nervios*) which is described as "a sudden partial loss of consciousness, accompanied by either clonic or tonic seizures and at times by screaming, tearing of clothing, or foaming at the mouth. Such *ataques* may last from a few minutes to as long as four days." These *ataques de nervios* have entered the medical literature as the "Puerto Rican syndrome" (Harwood 1981: 418).

Ataques de nervios are thought to be more common among women and characteristic of the lower socioeconomic classes. It may be culturally appropriate and expected behavior or may be an idiosyncratic response to an individual's life experience (Murillo-Mohde 1976; Padilla 1958). Garrison (1977) suggests that the *ataque* is a "culturally recognized, acceptable cry for help or an admission of inability to cope, and family and friends are required by norms of good behavior to rally to the aid of the *ataque* victim and relieve the intolerable stresses." It might appropriately occur in response to grief or receiving news of a shocking event, particularly when it is related to a loved one (ibid.).

Dresp's study (1985) reports on a series of interviews with seven Puerto Rican women suffering from *nervios* who were seen at a neighborhood clinic affiliated with Cambridge City Hospital. The women, aged 20 to 45, were asked 23 open-ended questions in Spanish both at the clinic and in their homes. The major finding was that all subjects reported that there was a specific precipitating event that was said to have caused the interviewee's *nervios*. Five respondents mentioned some kind of loss, either of a loved one or of a familiar environment, and the ensuing loneliness (Dresp 1985: 123). One case of *nervios* began after a car accident, and another with the woman's last month of pregnancy. The most prevalent sensation accompanying *nervios* was restlessness followed by shaking, feeling out of control, headache, and experiencing a bad mood or the urge to scream or cry (Dresp 1985: 125).

Koss-Chioino (1989) reports on 770 cases of women seen by medical doctors and mental health therapists for "anxiety disorder" of which the "base symbolic complex" is *nervios*/nervousness. Nerves was presented to therapists by 31 percent of the women and by 33 percent of the men. Associated problems include sleep disturbances, restlessness, headaches/ head problems, nervousness, crying jags, hostility, hallucinations, trembling, and bad humor. Koss-Chioino concludes that *nervios* is a culturally patterned symptom presented by many women who suffer anxiety disorders and reflects bodily malaise that disrupts thought and behavior and results in feelings of confusion and a lack of control (ibid.: 270).

The work of Peter Guarnaccia (Guarnaccia, DeLaCancela, and Carrillo 1989; Guarnaccia, Good, and Kleinman 1990; Guarnaccia, Rubio-Stipec, and Canino 1989) has focused on *ataques de nervios* among Puerto Ricans and other Latino migrants in the United States. According to their research, *ataque de nervios* is an acute episode during which the person shakes, feels heat or pressure in their chest, feels weakness in and/or difficulty moving their limbs, has a sense that his or her hands or face have "fallen asleep," their mind goes black and/or they lose consciousness. The person experiences no memory of what happened during the episode (Guarnaccia, DeLaCancela, and Carrillo 1989: 51). Symptoms vary from person to person for both *nervios* and *ataque de nervios* although there is general agreement that an *ataque* occurs from a build-up of *nervios* and some stressful life event acts as a trigger (Guarnaccia, DeLaCancela, and Carrillo 1989: 51). In one example, Señora Gomez describes her *ataque* as "shaking all over and not being able to control the shaking" (Guarnaccia, DeLaCancela, and Carrillo 1989: 53), while in another, Señora Ramos describes her experience differently: "I was sitting calmly in my house and watching television. Suddenly a heat inside began to rise to my face. I felt dizzy. I got cold. My heart began to beat strongly, as if it had a great weight on it. I felt sick all over" (Guarnaccia, DeLaCancela, and Carrillo 1989: 54).

Guarnaccia argues that *ataques de nervios* are provoked by abuses by family members, working hard for one's children and being abandoned by them, and disruption in both family and community networks often triggered by "the social stresses of poverty, migration, acculturation, discrimination, and unemployment" (ibid.: 60).

In another study, Guarnaccia, Rubio-Stipec, and Canino (1989) examine *ataque de nervios* in terms of symptom responses to the Puerto Rican Diagnostic Interview Schedule from a representative sample of 1513 cases on the island of Puerto Rico. They found that the *ataque* symptoms identified individuals who fit the social characteristics of *ataque* sufferers in the ethnographic literature. The 348 (23 percent) of the people in the *ataque de nervios* category were more likely to be female, over 45, much less likely to have finished high school and reported family incomes of under $10,000 per year (Guarnaccia, Rubio-Stipec, and Canino 1989: 286). Findings included that there is a close correspondence between panic symptoms and the symptoms of *ataque de nervios* and that the excess report of somatization symptoms is in part influenced by reports of *ataque de nervios*.

Nervios, then, for Puerto Ricans in the United States is most generally an embodiment of loss and family disruption, but as in the Guatemalan example, underlying sociopolitical distress produced by the conditions of poverty, migration, and discrimination can not be separated from their more personal consequences. Moreover, the metaphor of *nervios* as grieving seems to play a more important cultural role in the interpretation of the experience.

Eastern Kentucky

Any clinician working Appalachia is familiar with the complaint of "nerves." This common complaint tends to be used in several ways: as an independent condition, as a complication of another problem, or as an etiological explanation for other symptomatology. Surprisingly, this seemingly ambiguous complaint possesses substantial social and medical communication value among its users. Family members and neighbors nod their heads knowingly and sympathetically when the patient indicates that the person is suffering from this socially acceptable malady. Physicians themselves, when unable to make a specific diagnosis, readily resort to this catch-all label to account for a plethora of symptoms (e.g. "Nothing serious to be worried about, Mary Lou ... it's just your nerves."). (Ludwig and Forrester 1981)

The earliest studies of nerves in Eastern Kentucky have been reported in the literature by physicians and clinical researchers who are interested in the treatment of this complaint among their Appalachian patients. The data collected in these studies reflect this clinical bias which influences symptom reports and interpretation. Nonetheless, these studies describe the presentation of nerves and discuss its cause and folk etiology. More recently Eileen Van Schaik (1989) and Nations, Camino, and Walker (1989) have completed studies that present the cultural dimension of nerves in the Eastern Kentucky context. Both the clinical and ethnographic studies are presented in this summary.

In a psychiatric study of Eastern Kentucky coal miners between the ages of 40 and 55, Arny (1955) found a series of symptoms including fainting, partial paralysis, forgetfulness, and amnesia which were attributed to nerves. Physical complaints such as inertia and weakness and the inability to exert oneself without getting faint, shaky, or "going to pieces" were said to be caused by "worn out nerves."

Ludwig and Forrester (1981, 1982) were interested in delineating the nature of nerves by comparing it to symptomatic presentations which were more clearly somatic or psychiatric. They administered the Cornell Medical Index to 101 Social Security Disability petitioners scheduled for psychiatric evaluation at a rural Eastern Kentucky clinic. Nerves patients endorsed the greatest number of symptoms and scored the highest on the Cornell Medical Index. The complaints presented with nerves included "being nervous and shaky in the presence of superiors, getting rattled or confused easily during decision-making or under pressure, feeling keyed up and jittery, and being easily startled" (ibid.: 333). The most commonly reported symptoms were faintness, dizziness, hot and cold spells, sweating, and trembling; patients describe the trembling as being inside and not observable to others.

Van Schaik's (1989) more recent anthropological account is based on research conducted with eight women and two men ages 14 to 82 interviewed in a community clinic, a Veterans Administration Hospital, the University of Kentucky Medical Center, or in their homes. Nerves symptoms included feelings of nervousness, anger, impatience, fearfulness, and depression as well as agitation and restlessness, insomnia, and crying. "The one symptom that all informants report is some form of physical agitation of restlessness including shaking, trembling, fidgeting, jitteriness, itching, and internal quivering or jerking. In addition, individuals are unable to sit still, pace the floor when they cannot sleep, and must have something to do" (Van Schaik 1989: 19). All but one of the interviewees reported that family difficulties, compounded by financial worries, were the source of their nerves.

For Van Schaik, nerves embodies the experience of class inequality and domination – within, I would add, the United States that represents itself as an egalitarian society. Nerves, thus, is a metaphor of the contradictions inherent in being a member of the rural underclass while participating in the dominant ideology. The notion of "worn out nerves" suggests a capitalist production metaphor with the sufferer's body being the site of symbolic depletion and worthlessness.

Nations et al. (1989) on the other hand, suggest that nerves is a lay idiom for emotional distress and consider nerves a folk ailment related to anxiety and depression. In a study of 149 patients from University Medical Associates, 47 presented nerves. The most common physical symptom, reported by 87.2 percent of patients, is a "shaky, jumpy or jangly" sensation which manifests "everywhere on the inside," but "just under the skin," "on the outside" or more specifically in the hands, chest or stomach (ibid.: 1248). An acute attack is described as a wave of tension which breaks over the patient in a fit of body shaking and quivering, stomach pains and nausea, headaches, heart tremors, shortness of breath, chest pains, dizziness, trembling in the extremities, blurred vision, and hot flashes; it then calms into a state of lethargy, sadness, tension insomnia and feelings of weariness, of weight pressing on the chest, arms and head, and of obsessive worrying. Florence, a 40-year-old black woman originally from rural Virginia, suffering from nerves describes her symptoms as including "shaking, dizziness, constant worrying, loss of sexual feeling, sleeplessness, sharp knife-like stomach pains, weak knees, staggering, spinning head, and 'falling out' spells during which she falls to the floor, paralyzed and unable to speak" (ibid.: 1246).

Nerves was most commonly attributed to misfortune and tragedy and more frequently reported by women and housewives. The majority of patients never discussed their ailment with a physician. Those patients who were seen in the clinic were treated by being told that nothing is wrong or by referral to another clinic, or were prescribed medications such as Valium,

Librum or Benadryl (ibid.: 1255). In summary, Nations et al. state that nerves "is a powerful indicator of emotional distress and dysphoria" (ibid.: 1248).

Thus, for Nations et al., nerves embodies emotional distress. Its metaphorical message seems to focus on the "nervous" aspect of the experience, with the body experiencing trembling, shaking, and jittery feelings. Van Schaik and Nations et al. suggest that nerves is a powerful indicator of distress, but their evidence suggests that it has been heavily medicalized in the Kentucky situation, which transforms the basis of its metaphorical power.

Outport Newfoundland

The study of nerves in Outport Newfoundland is based on the work of Dona Lee Davis (1982, 1983, 1988, 1989) who has been conducting research in the southwest-coast fishing village of Grey Rock Harbour, composed of approximately 800 residents. Based on participant observation and a menopause questionnaire that was given to 38 women between the ages of 35 and 65, Davis found that 80 percent of the middle-aged women constantly talked about their nerves and their states of being in relation to their nerves (Davis 1983).

Davis reports that in local usage, nerves is a physical, psychological and cultural phenomenon. "Physically nerves are 'little strings that hold your body together.' Psychologically nerves are feelings of worry and fear. Culturally, nerves represent the role and moral duty to the fisherman's wife. Nerves may be brought on by worry or hard times, but you can also inherit nerves or learn nerves' behavior from others. Moreover, nerves are considered normal" (Davis 1982: 213). Dinham (1977) adds that nerves is a way of saying that one does not suffer from mental illness and acts as a socially acceptable expression of psychological stress. Antler (1980) relates nerves to the stress of the low social status of women in the rapid industrialization of the Newfoundland fishery.

Nerves, in its simplest form, is experienced "like being on pins and needles," or being "all a-jump" or "jittery" (Davis 1989: 68). The embodiment of nerves in Outport Newfoundland is experienced as normal, part of everyday life. Nerves is an embodiment metaphor of worrying, and worrying is seen as natural and normal in this fishing community where women wait for their husbands who are at sea. Yet it seems that the Newfoundland case introduces a moral dimension to the metaphor of nerves; women expect nerves and accept the suffering as part of being a good wife and a hard worker.

The embodiment of nerves

These *nervios*/nerves cases illustrate the pervasiveness of nerves being experienced as trembling, shaking, twitching of the body; unusual sensations of hot and cold and body aches; sense of disorientation and dizziness; insomnia, weakness, and debility; losing control, losing consciousness, fainting, and temporary paralysis; and not feeling like oneself or feeling outside of oneself. These sensations not only reappear from culture to culture, they also seem to describe varying "senses of the body." Table 6.1 summarizes a preliminary attempt to group together some of these sensations with regard to the expression of embodiment. These groupings describe bodily experiences that range from a sense of the body surface moving or the body shaking and incorporating foreign (not of the body) sensations to a loss or diminishing of the sense of body altogether. For instance, I interpret the experiential statements of "spinning" or "disoriented" as a disturbance of body perception and "insomnia" and "weakness" as a disturbance of body functioning. The most extreme examples of the experience of body disturbance may be when individuals feel as if they have "lost control" or fainted, in which the sense of body actually comes and goes; the disturbance is that the sense of body is unpredictable. Within this preliminary schema, statements like "I don't feel like myself" or "not acting like myself" may reflect a sense of body that is lost or unrecognizable, but may also be interpreted as a loss of self. The last example, "going to pieces" may also be interpreted in terms of both the sense of body, that is, the body is falling apart, or a sense of fragmentation of the self.

These "senses of body" may provide further clues to nerves as an embodied *metaphor* with meanings, not just sensations, that have some commonalities cross-culturally. The form and experience of *nerves* embodiment suggest that the bodily experiences are metaphors of self/society relations, with the body acting as the mediating symbolic device. While the cultural context remains, the locus of the specific symbolic language of the metaphors – the body shaking, incorporation, extrusion, loss of feeling, and disappearance – suggests a kind of metaphoric transparency to the vulnerability of the individual sufferers within the hostile and sometimes violent contexts of their lives (see also Jenkins and Valiente, Chapter 7 of this volume). Nerves embodiment, thus, can communicate the disintegration and breakdown of the self/society contract in explicitly bodily terms with the body disturbances reflecting the nature of that disturbance. The illness embodiment is metaphorically framed in symbolic terms that are specific to the cultural and social context. These embodied metaphors, therefore, can communicate the specifics of culture and the commonalities of the human condition.

Nervios/nerves is clearly the embodiment of psychological, social, poli-

Table 6.1. *Nerves as a disturbed sense of body*

Sensation	Sense of body	Culture

"body shakes"
"trembling,
  prickling"
"twitching of
  the body"
"jerking of
  the body"
	sense of the
	body surface
	or entire
	body moving
		Costa Rica
		Guatemala
		Kentucky
		Puerto Rico
		Newfoundland

"body aches,
  headaches"
"brain pain"
"hot and cold
  sensations"
"sweats"
"queer feelings
  in my head"
	body sense includes
	foreign sensations
		Costa Rica
		Guatemala
		Puerto Rico
		Kentucky
		Newfoundland

"was spinning" "disoriented" "dizziness" "feel faint"	body perception is distorted	Costa Rica Guatemala Kentucky
insomnia weakness debility	sense of body functioning is reduced	Costa Rica Newfoundland Kentucky

"lost control"
"blacked-out"
"lost
  consciousness"
"partially
  paralyzed"
"fainting"
	sense of body comes and
	goes; loss of sense of body
	for a short time
		Costa Rica
		Puerto Rico
		Kentucky

"don't feel
  like myself"
"not acting
  like myself"
"feel outside
  of myself"
"being
  temperamental"
	sense of body is not there,
	or is so changed that it is
	not recognizable
		Costa Rica
		Puerto Rico
		Kentucky
		Newfoundland

"going to pieces"	sense of body falling apart	Costa Rica Puerto Rico Kentucky Newfoundland

tical, and cultural distress. It has been described as embodying everything from family disruption and personal loss to political terrorism and class inequality. The cases illustrate rather effectively the contention that nerves is the embodiment of all adverse existential conditions and disorder in the world. Nerves, however, as an embodied metaphor also carries the communicative force of culturally generated metaphors of distress that provide symbolic expression of personal conflicts, community upheaval, and social control through bodily experience.

So we return again to the initial dilemma. The self/society suffering communicated through the embodied metaphor of nerves becomes part of the domain of medicine. If the expression were otherwise, that is, if it were expressed with violence or rage, it might become part of the criminal or political rather than the medical world. So even embodied metaphors that retain the holism of experience may medicalize experience and thus limit the potential responses and behaviors elicited.

ACKNOWLEDGEMENTS

Sections of the case studies presented were previously published in Low (1985). I would like to thank Carole Browner, Laurel Wilson, Lucile Newman, Joel Lefkowitz, and Thomas Csordas for their helpful comments. I also would like to thank my colleagues, Mary Parlee and Rachel Brownstein, and the graduate students who participated in the "Conceiving the Body" seminar at the City University of New York Graduate Center.

REFERENCES

Antler, E. (1980) Newfoundland Women: The Economics of Home Economics. *Anthropological Resource Center Newsletter* 4: 4–6.

Arny, M. (1955) My Nerves are Busted. *Mountain Life and Work* 3: 24–9.

Barlett, P. and Low, S. (1980) Nervios in Rural Costa Rica. *Medical Anthropology* 4: 523–64.

Cayleff, S. (1988) Prisoners of their own Feebleness: Women, Nerves and Western Medicine. *Social Science and Medicine* 26: 1199–209.

Csordas, T. (1987) Genre, Motive and Metaphor. *Cultural Anthropology* 2: 445–69.

(1990) Embodiment as a Paradigm for Anthropology. *Ethos* 18: 5–47.

Davis, D. L. (1982) Woman the Warrior: Confronting Feminist and Biomedical Archetypes of Stress. Presented at the 81st annual meeting of the American Anthropological Association, December 3–7.

(1983) *Blood and Nerves*. St John's: Memorial University of Newfoundland Institute of Social and Economic Research.

Davis, D. (1988) Introduction. Historical and Cross-Cultural Perspectives on Nerves. *Social Science and Medicine* 26: 1197–9.

(1989) The Variable Character of Nerves in a Newfoundland Fishing Village. *Medical Anthropology* 11: 63–78.

Davis, D. and P. Guarnaccia (1989) Health, Culture and the Nature of Nerves. Special issue of *Medical Anthropology* 11: 1–95.

Davis, D. and S. Low (1989) *Gender, Health and Illness*. Washington DC: Hemisphere Publishing.

Dinham P. S. (1977) *You Never Know What They Might Do*. St John's: Social and Economic Studies No. 20. Institute of Social and Economic Research Memorial University of Newfoundland.

Dresp, C. W. (1985) Nervios as a Culture-Bound Syndrome among Puerto Rican Women. *Smith College Studies in Social Work* 55: 115–36.

Dunk, P. (1989) Greek Women and Broken Nerves in Montreal. *Medical Anthropology* 11: 29–45.

Fernandez, J. (1986) *Persuasions and Performances*. Bloomington: Indiana University Press.

Finkler, K. (1989) The Universality of Nerves. *Health Care for Women International* 10: 171–9.

Foucault, M. (1979) *Discipline and Punish*. New York: Vintage.

Frank, G. (1986) On Embodiment. *Culture, Medicine and Psychiatry* 10: 189–220.

Garrison, V. (1977) The Puerto Rican Syndrome in Psychiatry and Espiritismo. In V. Crapanzano and V. Garrison, eds., *Case Studies in Spirit Possession*, New York: Wiley.

Good, B. (1990) Narrativity and Suffering. Paper presented at the American Anthropological Association annual meeting, November, New Orleans.

Gordon, D. (1990) Embodying Illness, Embodying Cancer. *Culture, Medicine, and Psychiatry* 14: 275–97.

Guarnaccia, P., V. DeLaCancela, and E. Carrillo (1989) The Multiple Meanings of Ataques de Nervios in the Latin Community. *Medical Anthropology* 11: 47–62.

Guarnaccia, P., M. Rubio-Stipec, and G. Canino (1989) Ataques de Nervios in the Puerto Rican Diagnostic Interview Schedule. *Culture, Medicine and Psychiatry* 13: 275–96.

Guarnaccia, P., B. Good, and A. Kleinman (1990) A Critical Review of Epidemiological Studies of Puerto Rican Mental Health. *American Journal of Psychiatry* 147: 1449–56.

Harwood, A. (1981) Puerto Rican Americans. In A. Harwood, ed., *Ethnicity and Medical Care*. Cambridge, MA: Harvard University Press, pp. 418–61.

Hill, C. and L. Cottrell (1986) Traditional Mental Disorders in a Developing West Indian Community in Costa Rica. *Anthropological Quarterly* 59: 1–14.

Jenkins, J. (1988a) Conceptions of Schizophrenia as a problem of Nerves. *Social Science and Medicine* 26: 1233–44.

  (1988b) Ethnopsychiatric Interpretations of Schizophrenic Illness: The Problem of *Nervios* within Mexican American Families. *Culture, Medicine and Psychiatry* 12: 301–30.

  (1991) The State Construction of Affect. *Culture, Medicine, and Psychiatry* 15(2): 1–39.

Johnson, M. (1987) *The Body in the Mind*. Chicago: University of Chicago Press.

Johnston, F. and S. Low (in press) *Children of the Urban Poor: Child Development and Malnutrition in Guatemala City*. Boulder: Westview Press.

Johnston, F., S. Low, Y. Baessa and R. MacVean (1985) Growth Status of Disadvantaged Urban Guatemalan Children of a Resettled Community. *American Journal of Physical Anthropology* 68: 215–24.

Kirmayer, L. (1988a) Mind and Body as Metaphors. In M. Lock and D. Gordon, eds., *Biomedicine examined*. Dordrecht: Kluwer Academic Publishers, pp. 57–94.

(1988b) The Body's Insistence on Meaning. Paper presented at the XII International Congress of Anthropological and Ethnological Sciences, Zagreb, July, pp. 3–40.

(1992) The Body's Insistence on Meaning: Metaphor as Presentation and Representation in Illness Experience. *Medical Anthropological Quarterly* 6: 323–46.

Koss-Chioino, J. (1989) Experience of Nervousness and Anxiety Disorders in Puerto Rican Women. *Health Care for Women International* 10: 245–72.

Lakoff, G. (1987) *Women, Fire and Dangerous Things*. Chicago: University of Chicago Press.

Lock, M. (1988) A Nation at Risk: Interpretations of School Refusal in Japan. In M. Lock and D. Gordon, eds., *Biomedicine Examined*. Dordrecht: Kluwer Academic Publishing, pp. 377–414.

(1990) On Being Ethnic: The Politics of Identity Breaking and Making in Canada, or *Nevra* on Sunday. *Culture, Medicine and Psychiatry* 14: 237–54.

Lock, M. and P. Dunk (1987) My Nerves are Broken. In D. Coburn, C. D'Arcy, G. Torrance and P. New, eds., *Health and Canadian Society: Sociological Perspectives*. Toronto: Fitzhenry and Whimbeside, pp. 295–313.

Low, S. M. (1981) Meaning of *Nervios*: A Sociocultural Analysis of Symptom Presentation in San José, Costa Rica. *Culture, Medicine and Psychiatry* 5: 25–47.

(1985) Cultural Interpreted Symptoms or Culture-Bound Syndromes: A Cross-Cultural Review of Nerves. *Social Science and Medicine* 21: 187–96.

(1988) Medical Practice in Response to a Folk Illness: The Treatment of *Nervios* in Costa Rica. In M. Lock and D. Gordon, eds., *Biomedicine Examined*. Dordrecht: Kluwer Academic Publishers, pp. 415–40.

(1989) Gender, Emotion, and *Nervios* in Urban Guatemala. *Health Care for Women International* 10: 115–39.

Ludwig A. M. and R. L. Forrester (1981) The Condition of "nerves." *The Journal of the Kentucky Medical Association* 79: 333–6.

(1982) ... Nerves, but not Mentally. *Journal of Clinical Psychiatry* 43: 187–90.

Martin, E. (1987) *The Woman in the Body*. Boston: Beacon Press.

Mauss, M. (1950) Les Techniques du Corps. *Sociologies et Anthropologie*. Paris: Presses Universitaires de France.

Merleau-Ponty, M. (1962) *Phenomenology of Perception*. New Jersey; Routledge and Kegan Paul.

Murillo-Mohde, I. (1976) Family Life among Mainland Puerto Ricans in New York City. *Perspectives in Psychiatric Care* 14: 174–9.

Nations, M., L. Camino and F. Walker (1989) "Nerves": Folk Idiom for Anxiety and Depression. *Social Science and Medicine* 26: 1245–59.

Quinn, Naomi (1987) The Cultural Meaning of Metaphor. Paper presented at the American Anthropological Association, Washington DC.

Padilla, E. (1958) *Up from Puerto Rico*. New York: Columbia University Press.

Pandolfi, M. (1990) Boundaries Inside the Body: Women's Sufferings in Southern Peasant Italy. *Culture, Medicine and Psychiatry* 14: 255–74.

Sacks, O. (1984) *A Leg to Stand On*. New York: Harper and Row.

Scarry, E. (1985) *The Body in Pain*. New York: Oxford University Press.

Scheper-Hughes, N. (1989) Bodies, Death and the State in Northeast Brazil. Paper presented at the American Anthropological Association annual meeting, November, Washington.

(1992) *Death Without Weeping*. Berkeley: University of California Press.

Scheper-Hughes, N. and M. Lock (1986) Speaking "Truth" to Illness: Metaphors, Reification and a Pedagogy for Patients. *Medical Anthropology Newsletter* 17: 137–9.

(1987) The Mindful Body. *Medical Anthropology Quarterly* 1: 6–41.

Sluka, J. (1989) Living on their Nerves: Nervous Debility in Northern Ireland. *Health Care for Women International* 10: 219–44.

Taussig, M. (1987) *Shamanism, Colonialism, and the Wild Man: A Study in Terror and Healing*. Chicago: University of Chicago Press.

Turner, B. (1984) *The Body and Society*. London: Basil Blackwell.

Van Schaik, E. (1989) Paradigms Underlying the Study of Nerves as a Popular Illness Term in Eastern Kentucky. *Medical Anthropology* 11: 15–28.

7   Bodily transactions of the passions: *el calor* among Salvadoran women refugees

Janis H. Jenkins and Martha Valiente

Over the course of the last decade, there has been a paradigmatic shift in anthropological thinking on the construct of "emotion." No longer assumed to index a fundamentally universal, psychobiological event for individuals, emotion is currently theorized as culturally constituted and situationally specific to social realms (Lutz 1988). The largely unanalyzed convergence between Western scientific and popular views of emotion was until recently not noted as particularly suspicious. Indeed, previous universalist–individualist accounts of emotion are now construed by many as but one ethnopsychological creation myth (Abu-Lughod and Lutz 1990; Kleinman and Good 1985; Lutz 1985; White and Kirkpatrick 1985). Feminist theories have also deconstructed the ideology inherent in symbolic representations of emotion within dichotomous realms of the devalued natural, dangerous, and female, on the one hand, and the more esteemed cultural, controlled, and male, on the other (Haraway 1991; Ortner 1976; Rosaldo 1984; and especially Lutz 1988, 1990). Not surprisingly, the new emphasis on the sociocultural construction of emotion has occasioned a wave of cultural studies in this area (Abu-Lughod 1986; Desjarlais 1992; Gaines and Farmer 1986; Good and Good 1988; Jenkins 1991b; Kleinman 1986; Lutz and White 1986; Markus and Kitayama 1994; Matthews 1992; Myers 1986; Ochs and Schieffelin 1989; Rosaldo 1980; Roseman 1991; Scheper-Hughes and Lock 1987; Schieffelin 1976; Shweder and LeVine 1984; Wikan 1990; see also Lyon and Barbalet, Chapter 2 of this volume). In a celebratory mood, anthropologists echo Geertz's (1973: 81) observation that "not only ideas, but emotions too, are cultural artifacts."

A consequence of culturalist approaches to emotion, however, has been the estrangement of culture from the body in the name of anti-reductionism.[1] Only recently have notions of the body as a generative source of culture, experience, and orientation emerged alongside more cognitive interest in mental representations such as knowledge, symbols, and meanings as the presumed loci of culture. Long-standing dualisms of the mind as cultural and the body as biological have often served to render the physical, sensational world of pangs, vapors, and twinges theoretically insignificant and largely absent from cultural-symbolic analysis.

While some anthropologists and psychologists (e.g. Levy 1984 and Frijda 1987, respectively) suggest an operational separation of "feeling" (sensation) and "emotion" (cognized interpretation), we prefer here to problematize this distinction from the vantage point of the body. From this perspective, we wonder about the heralding of "complex" emotions such as *amae* (Doi 1973) at the expense of "simple" sensationally and bodily based emotions such as "angry livers" (Ots 1990). We wonder whether this line of thinking is predicated on the traditional dualist idea that the closer we come to the body the farther away we must be from culture. As will become more vivid following our ethnographic discussion, we intend our essay as a critique of conceptualizations of the body as a *tabula rasa* upon which culture inscribes its codes. Rather, we are impressed with the degree of intentionality and agency of the body in creating experience. Although it is possible to access such worlds through analyses of mental representations such as language and ethnopsychological knowledge, the cultural creation of intersubjective realms of social space via the body has often eluded the anthropological gaze.

Another orienting premise of this essay is that social domains of power and interest are constitutive of emotional experience and expression (Corradi et al. 1992; Good et al. 1988; Jenkins 1991a; Lutz and Abu-Lughod 1990; Kleinman 1986; Scarry 1985; Swartz 1991). While recognition of the essential interrelations between the personal and the political has long been central to feminist scholarship (Rosaldo and Lamphere 1974), this point has yet to be adequately integrated in theories of culture and emotion. To this end, Abu-Lughod and Lutz (1990) have recently proposed new theoretical directions for analyzing sociopolitical dimensions of emotion in everyday discourse. Scholarly discourse on the emotions has also been considerably expanded by Good and Good (1988) through conceptualizations of the "state construction of affect." Jenkins (1991a: 140) has urged that the emerging scholarly discourse on the emotions include the nexus of the role of the state in constructing a *"political ethos"* and the personal emotions of those who dwell in that ethos.[2] Suarez-Orozco (1990: 353) proposes examination of the formal structures or "grammar" of collective terror such as is now widespread in Latin America.

In this chapter, we are concerned with the cultural and sociopolitical basis of bodily experience.[3] Our interest lies not merely with the sociopolitically "inscribed" body but also with the body as seat of agency and intentionality through resistance, denial, reactivity (Shweder 1990; Scheper Hughes 1992). We present an ethnographic analysis of a particular form of bodily experience – *el calor* (the heat) – among Salvadoran women refugees seeking help at an outpatient psychiatric clinic in North America.[4] Our analysis of the narratives of Salvadoran women is presented (1) as documentation of a culturally specific form of bodily experience that is relatively unknown in North American medical settings; and (2) as an empirical basis for theorizing on the interrelations of culture, emotion, and the body. The argument is intended as a contribution to discourse on the state construction of affect, on the one hand, and of the intersubjectivity of those affects, on the other.

The present study of Salvadoran refugees is based upon the clinical and narrative presentations of twenty-two women living in a metropolitan area of the northeastern United States. The study was ethnographic, including semi-structured and informal interviews, and participant observation in home, community, and clinical settings. Each woman was encountered in the process of seeking or receiving help from an outpatient psychiatric clinic at a university teaching hospital. At the time of contact with the hospital, nearly all women reported symptoms of major affective or anxiety disorders, especially including major depressive and post-traumatic stress disorders.[5] Most of the women had been in the United States for at least one year and had family, including young children, who were still residing in El Salvador. Many worked long hours – sixty or more in two jobs – in vigorous efforts to make as much money as possible to send back home to relatives.

Escape from *la situación*

Refugees' narratives of their emigration from El Salvador often highlight escape from *la situación*.[6] *La situación* is a rhetorical term that implicitly, and possibly covertly, refers to unrelenting political violence and poverty from which these women have fled (Jenkins 1991a). Although violence and civil warfare have been common in El Salvador throughout this century, the most intensive sustained conflict to plague the country occurred in the period of 1979–92. Since 1979, the wave of warfare and terror has decimated the population by death and emigration. At least 75,000 persons were killed during this period, with several thousands more *"desaparecidos"* (or disappeared), 500,000 internal refugees, and an estimated 1,000,000 more who fled to other countries such as Mexico, Honduras, Panama, the United States, and Canada. In a country that, in 1979, had a population estimated at 5.2 million persons (Fish and Sganga 1988), the decimation of the Salvadoran people becomes all too evident.

In various combinations, the women in the present study gave three primary reasons for their flight from *la situación* in El Salvador: escape from large-scale political violence, escape from "domestic" violence,[7] and escape from impoverished economic conditions. These practices of the state, the economic conditions, and the domestic environment are appropriately understood not as independent factors but as coordinate dimensions of a single political ethos. Martin-Baro (1990) has characterized the entire nation as one in which state induction of fear, anxiety, and terror is elaborated and maintained as a means of social control. Through long-term exposure to this political ethos the experience of the "lived body" is dominated by anxiety, terror, and despair (Jenkins 1991a: 149). Regular encounters with the brutality of the war were commonplace: mutilated bodies along the roadside or on the doorstep to one's home, disappeared loved ones, the terror of military troops marching through town and shooting at random and arresting others who would be incarcerated and tortured (Jenkins 1991a). Life within this landscape of political violence compelled these Salvadorans to flee in search of a safe haven (see also Farias 1991).

The stirring of bodily memory: an experiential analysis of *el calor* (the heat)

For nearly all of the women in the study, flight from political violence is narrated within the context of problems with *nervios* (nerves). *Nervios* refers at once to matters of mind, body, and spirit and does not make cultural sense in relation to mind–body dualisms. While the cultural meaning and specific symptom profile of the indigenous condition of *nervios* varies across cultural settings (Low 1985, 1988, Chapter 6 of this volume), the cultural category appears widely throughout Latin America. *Nervios* indexes a broad spectrum of distress and illness and includes everything from minor situational upsets to an established schizophrenic illness (Jenkins 1988).[8] *Calor* (heat) has sometimes been included as a component or symptom of *nervios* as a condition as well as of episodic *ataques de nervios* (Guarnaccia et al. 1990).

Within the context of El Salvador, narratives of *nervios* are deeply embedded within the life situation of chronic poverty and exposure to violence. Among these women, *calor* is but one among several bodily phenomena associated with *nervios*. Other bodily sensations include *escalofrios* (shivers, chills), *un hormigueo en la piel* (sensation of a swarm of ants on the skin), *un adormecimiento* ("sleepiness" or numbness) localized on one side of the face or body, *choques electricos* (electric shocks), and feelings of being *inquieta* (agitated), often with *ganas de correr* (the urge to run). Because reports of *el calor* emerged for the majority of women and because these are problematic for anthropological and medical understandings of culture, emotion, and the body, we choose to examine it here. The remainder of this section of the chapter is organized into a discussion of (1) narratives of *el calor*; (2) ethnographic accounts of the perceived bodily sites and degrees of the severity of *calor* (3) tropes and other linguistic conventions for describing *calor*; and (4) the social situational and emotional contexts of *calor* experiences.

Narratives of calor

Accounts of personal experience with *calor* were either spontaneously offered during the course of the interviews or were given in response to direct queries.[9] Excerpts from the narratives provide evidence of the polysemous, multivocal nature of discourse on *el calor*:

1. From Gladys Gonzalez, a 39-year-old woman living with her husband and five children:

Heat is something like fire that you can feel from toe to head ... throughout your body ... nothing more than a vapor that you feel and then it passes ... a hot welling up.

2. From Adelina Valenzuela, a 56-year-old Salvadoran living with her daughter, grandchildren, and three other extended family members:

I used to feel hot currents rising up in inside of me, hot, I felt like I was suffocating, and that despair that comes when the heat rises up, something hot, I couldn't [resist it] and I didn't feel well and my vision blurred ... [the heat] rose upward inside from my feet to my head throughout my body ... throughout my body, I felt like the fire was shooting out of me, from here and here and my eyes felt like they were being pushed out by the fire, like they were going to come right out, and from the ears I felt it coming out of me ... and in my mouth I could begin to taste ... vapor, my own breath. I felt like it was fire, a flame, and it was inside of me ... I was desperate, the heat, a terrible thing ... I felt like I was suffocating and that I was dying [and] I went and turned on the cold water to take a cold shower ... the heat feels as if, you know, with a match [you light] a sheet of paper and then swallow it, and inside the heat that feels so terrible, those flames of fire welling up.

3. From Elsa Hernandez, 36 years old, whose family still resides in El Salvador, and who lives with a Euro-American couple as a maid and caretaker for their children:

Calorias [Heat attacks], that's what we call them here in El Salvador, my mother also had them and they say it has to do with the blood, apparently it becomes irritated and well, my hands and neck become ... as if I had fever and I get real hot, but only in my hands and neck ... it's a type of heat, yes, that they call "the urge to bathe," yes, it doesn't matter what time, the heat you feel, even though it's cold ... it comes from worries, more than anything worries about important things, one is startled by one's nervous system and blood that are stirred by bodily memories. ✓

4. From Dora Campos, a 30-year-old woman separated from her *compañero* (partner) after repeated violent assaults upon her:

Heat, like some kind of vapor that rises upward from the feet, I don't know, I feel hot, I'm not sure how to explain what it's like. As if it begins in the feet and moves up until it reaches my head, at least that's what it was like before, but it's been a while since it happened ... [it happened] when I was back in my country, and also here a few times, but more back home – from time to time, I would feel bad and my body would feel like it was getting hot, as if ... I were to suddenly become very chilled [but] with sweaty palms, I think that it could be because of all of the problems I had that I couldn't vent, and the only thing I did was cry, and that didn't do me any good, also I didn't want people to see me crying, and I think that was the cause of all of this, the fact that I had so much pent up inside of me which I wasn't able to vent ... from the feet upward, something hot that made my face and head feel hot and then it dropped down and my hands began to sweat [and shake] nervously.

Bodily sites and severity of calor

El calor refers to intense heat which suddenly pervades one's whole body. Some report a particular body site, such as the head (face, ears, nose, and mouth, including taste and breath), neck, back, leg, stomach, chest, and hands, as an intense "centering point" from which it emanates. The onset of *calor* is rapid. It may commence in the feet, progressively intensifying and rising to the head. Other experiential accounts, however, describe *calor* as beginning in the head or neck and then spreading through the whole body. Although the experience is perceived as occurring *inside* one's body, it is thought to originate from without. It may be fleeting (a few minutes), or sustained (several days).

While some of the Salvadoran women affirm only occasional experience of *el calor*, others cite their experiences with it as too numerous to count. The range in this study was from four or five episodes to daily occurrences. The frequency and severity appear interrelated: the greater the number of episodes, the more intensely dysphoric the experience. Thus while some women said that they had known *calor* only a few times and in ways that were inconsequential and less than debilitating, others' more frequent episodes were recounted as virtually unbearable.

Thus the seriousness of *calor* varies considerably: for some it is perceived as a mild occurrence that is a normal part of everyday experience while for others it is experienced as a frightening event of potentially mortal consequences. In the fourth example cited above, Dora Campos recalled that following her relatively inconsequential *calor* experience she resumed her daily household activities without difficulty. In Adelina Valenzuela's (second example above) narrative, *calor* was described *como que era fuego* (as though she were fire), and the experience as a whole in terms of a dissociative state: *yo me sentía que no era yo* (I felt that I wasn't me). So terrifying was this *calor* experience that she feared she would surely die:

I felt a burning flame ... I felt I was dying, I felt an agony, something, I felt death was just above, with that heat.

In her frequent encounters with *calor*, she would typically feel compelled to take off her clothes, shower in cold water, and consult with both a psychotherapist and a *santero*[10] for healing. During one especially bad episode in which she thought she might asphyxiate, she ran out of her apartment and into the street. Upon recovery she remembered nothing. This particular case example falls within the indigenously defined cultural category of *ataques de nervios*, more fully described among Pureto Ricans by Guarnaccia, Kleinman, and Good (1990).

Tropes of calor

Vapor (vapor), *corrientes* ("electrical currents/surges), *fuego* (fire) or *llama* (flame) are common tropes for describing *el calor*. *Un vapor* (a vapor) is a steadily rising sort of "steam heat" that is felt throughout one's body. Although qualitatively intense, the experience nonetheless may have an insubstantial, fleeting quality that ultimately leaves the body, "evaporating" as would a steam heat. *Vapor* thus represents *calor* as a kind of incarnate substance. The sometimes "electric" movement of *calor* is captured by *los corrientes* (currents, surges) flowing through the body. This sense of fluidity may also be expressed as waves of fire or flame: "the heat is like fire in your whole body." Another woman described *calor* as the sensation of rolled up newspapers that were set ablaze in her chest. Yet another described it as *un vapor* like *el aliento* (the breath), felt as *un fuego* (a fire), or *una llama* (a flame) causing heat inside her body. In still other cases in her clinical practice, Valiente has observed that *calor* can also be colloquially communicated as *un fogaz* – a flame that shoots up through the body. Finally, the experience was also occasionally referred to as *calorias* (*calories, heat units*) or even *ganas de banarse* (the urge to take a bath). The women's attempt to communicate the phenomenon of *el calor* raises the perennial problem of the relation between language and experience. There is little doubt that *calor* is a cultural phenomenon for Salvadorans, but it is only a partially objectified one. We are constantly shifted among what appear as direct description, simile, and metaphor. Is *calor* itself a direct description, or a metaphor of an emotional state? Different informants may use the same word, for example *vapor*, as a simile (expressing similarity) or as a metaphor (expressing shared essence) for *calor*. In one of the examples just cited, we found *calor* to *be* a vapor that was *like* a breath *felt* as a flame *causing* heat.

Fernandez (1986) has given us the notion of the play of tropes, and Friedrich (1991) the term polytrope for such examples of metaphor upon metaphor upon simile. Friedrich notes that all tropes contribute to both ambiguity and disambiguation, and paraphrases Tyler to the effect that "a trope may mislead in exact proportion to the amount it reveals, but that is the price of any revelation" (1991: 24). Our example of *el calor* suggests not so much the masking of experience by linguistic representation as the indeterminate flux of bodily existence. The indeterminacy of these tropes reveals them not so much as cultural meanings imposed on experience as fleeting, evanescent disclosures of inexhaustible bodily plenitude. Metaphor and simile emerge from this plenitude in, to borrow Kirmayer's phrase, the body's insistence on meaning, which is "to be found not primarily in representation but in presentation: modes of action or ways of life" (1992: 380). This tropic movement is best described not in terms of Fernandez's (1986) dimension of inchoate to choate, but as a movement from preobjective indeterminacy to inexhaustible semiosis (cf. Daniel, Chapter 10 of this volume).

In addition to the reliance on simile and metaphor to partially communicate a largely incommunicable bodily experience, this indeterminacy is evident in the women's linguistic confusion over how best to refer to *calor*. Quite in contrast to their use of the culturally salient category of *nervios* (Jenkins 1988; Low 1985, Chapter 6 of this volume; Guarnaccia and Farias 1988), many fumbled or varied in the use of a definite or indefinite, feminine or masculine article prior to the nominal *calor*. For this reason, we inquired about their knowledge of *calor* as "*lo que le llaman*" (what they call) *el calor* or *la calor*. To be grammatically correct, the masculine "*el*" would be preferred. Nevertheless, our informants disagreed even on this point. Some informants readily employed and applied the term *el calor* to their experiences as described above, but a few claimed to know little of the term yet went on to describe the experience itself in ways that were relatively indistinguishable from those who did. These considerations suggest that while *calor* is a *cultural* phenomenon in so far as it is by no means commonly reported cross-culturally, it remains only *partially objectified* in the experience of Salvadorans. Let us follow the trail of meaning of this shadowy phenomenon farther into the lifeworld of the Salvadoran women refugees.

Social situational and emotional contexts of calor

Emotions associated with *calor* are nearly all dysphoric: *miedo, temor, susto,* and *preocupaciones* (fear, dread, fright, and worry); *desesperacion* (despair, desperation); *agonía y muerte* (misery, death agony); and *coraje, enojo, enfado* (anger). All these diverse affects are strong, to be sure, and were generally mentioned in relation to forceful experiences of *calor*. More mild experiences generally were not accompanied by this particular vocabulary of emotion. Thus experiences perceived to be "lighter" in character were noteworthy for the emotion words they failed to evoke in narrations. The possibility also exists, however, that some women simply eschewed specific mention of emotional terms on the grounds that these were unnecessary, inappropriate, or even unthinkable.

During the narrations of *calor* experience, some women provided examples of specific contexts, often the last time in which it had occurred. A few of these follow:

1. From 39-year-old Gladys Gonzalez, living with her husband and five children:

Before we were in El Salvador when my mother-in-law died and they called me to give her some water when she was dying. I felt my body very hot and numb but I had to face up to it. Then a neighbor was there too in those days. I had felt it too, because I had a nine year old son with a broken leg. I told him to get up but then I saw that he was in pain – he had lost his color – and that his heart hurt from the force he exerted with the crutches and just at that moment I felt a hot surge. I said to myself "he died."

In this example, Sra. Gonzalez provides two intimately intertwined family contexts of *calor* experienced as intense fear of actual, and imminently imagined, death. Faced with the real or potential loss of a family member, she responds bodily with what she thinks of as *el calor*.

2. From 35-year-old Lucrecia Canas, married mother of two, who still reside in El Salvador with her mother:

(Take what happened yesterday). I dropped a casserole dish with dinner in it and then nerves came on because my husband was right in front of me. When I dropped the casserole dish it gave me a shiver throughout my body and I felt immediate pain and then, so my husband wouldn't see that I was afraid, I didn't say anything. ... he had seen I dropped the dinner and since he is really angry, well, I, so he would see that I'm not afraid I said nothing to him. I had the heat attack in the moment I dropped the dinner. I felt an electrical charge was put inside my body. It was because of the fear I have of him, it's because he would have hit me at that moment, he would have beat me because I dropped the food.

Another example of Lucrecia's experience with *calor* as intensely embodied fear:

When he goes out drinking on Fridays, he comes back at three in the morning on Saturday, then I feel my face is on fire, really numb, the middle [of my face] only, and the agitation in my chest, I feel desperate, with an urge to leave [the apartment] running, and running, running to get far away ... I feel the desire to run away, but I don't actually do it [just as] there's the same pressure when he goes out drinking and returns irritable. Then I want to focus my attention and not be afraid of him, be strong, but I can't.

And a final example of embodied heat and *susto* (fright) in response to *la situación* prior to fleeing El Salvador:

In my country I had *un susto* (a fright) when a man was dying. Already the man couldn't speak [but] he made signs to me with his eyes. It was during the daytime, and I was going to get some chickens for a Baptism. He could barely move his eyes. He had been shot in the forehead. It was the time of the fair in November. When I came back he was already dead. I returned home with a fever, a great fever, and it wasn't something I'd ever experienced. Since it was carnival time, strangers came. They kill strangers.

3. From 47-year-old Reina Torres, married mother of three:

It happens to me if I feel bored, or it happens to me most when I am walking to the store with my husband, in the store, because with him it's boring to go out. Because nothing entertains him, nor does he say anything to me if I buy something or if something fits me well. He never says anything to me and then all of a sudden it grabs me, my leg goes to sleep, almost a side [of my body] like painful or my knee goes out, or my ankle.

This somewhat darkly amusing example of the body taking itself literally by actually "going to sleep" of sheer boredom is unique in the *calor* narratives. In terms of the emotional context of *calor* experiences, it fits neither with so-perceived prosaic, unemotional occurrences of mild impact nor with intense emotional experiences of anger or fear. Interpretively it might only be assimilated to one of these types if we were to consider it an example of angry boredom.

4. A powerful example of anger associated with heat, not part of a *calor* experience *per se*, was generated by Dora Campos, a 39-year-old mother of three who was presently torn over whether to separate from her physically assaultive husband:

[I]t is as if your blood . . . it's like putting water in a pot that's being heated and then letting it boil, it's as if your blood were boiling, and I feel as if I want to wreck everything [the world] if it were possible, but at the same time I can control myself a bit, because when my world is dominated by anger I prefer not to say anything [to hurt others] because I don't want to use angry words to express or reveal myself; I never know if such reactions are good or bad, for example when my husband becomes violent and mistreats me I fill up with anger [I become rabid], I feel angry but at the same time I overcome it and I say nothing, and if he comes up to me and threatens to hit me, I tell him to go ahead, and it is this manner which I believe allows me to control my anger.

Political and medical objectifications of the *calor* experience

Emotional orientations to cultural and sociopolitical realms

How might these diverse accounts of bodily experience be interpreted? *Calor* is an ethnographic example from a phenomenological world neither recognizable nor widely shared across cultural groups.[11] Within the realm of personal experience, *calor* can be conceived as a form of emotional engagement with social and political realities. Specifically, it is a somatic mode of attention (Csordas 1993, 1994), a mode of attending to and with the body in an intersubjective environment. Cultural variation in the elaboration or suppression of such somatic modes of attention is a potentially valuable dimension for examining cross-cultural differences in experiential and communicative worlds of emotion. We must insist that it would be a mistake to regard concepts such as *calor*, that only point to or outline a mode of attention, as sorts of bodily analogs of elements in the conceptual "belief system" of traditional cultural anthropology.[12] As Ots (1990: 22) has observed, "[t]his problem becomes even more acute when the concepts of others are not just a different mode of thought but can be viewed as created by a different kind of bodily perception, e.g., the concept of *qi* in Chinese culture."

In this context, difference in bodily perception refers not to the observation that different cultures tend to locate emotions in particular body sites or organs. The problem with organ-specific descriptions of emotion has been a failure to link body (or body part) to the social world, such that in fact they remain modeled on an intrapsychic conception of emotion (Lutz 1988). Ots (1990) considers bodily organs such as the "angry liver, the anxious heart, and the melancholy spleen" as evidence of the body's role in generating culture. In a critique of contemporary models of somatization (which he views as relegating bodily processes to psychological mechanisms presumed to be of a higher order), he argues that bodily manifestations can be considered as correspondents or equivalents of emotion (ibid.: 24; see also Ots, Chapter 5 of this volume; on pain as an emotion see Jackson, Chapter 9 of this volume). This intriguing idea repositions the construct of emotion within the lived body – quite the opposite of the more typical psychosomatic strategy that describes transformation of an essentially psychological event into a secondary somatic expression.[13]

Along these same lines, we can understand that *calor* is existentially isomorphic with anger and fear, with variations in the degree to which it is configured as primarily anger or fear or a thorough admixture of these. One significant contextual basis for these emotions is personal encounters with violence: violence of male kin and the immediate conditions of civil war represented as *la situación*. Each of these contexts contributes to the political ethos of a culture of terror in which brute violence is regularized. Cultural proscriptions of outwardly directed verbalizations of anger and rage by women are of obvious importance here.[14] Our analyses of refugees' narratives revealed that certainly not all or even most explicitly associated their experiences with either anger or fear. *Calor* may actively engage unjust worlds of violence through justifiable anger, but may also reactively engage these same worlds through fear and trembling. Personally and culturally unwelcome, the anger and fear that construct *calor* experiences engage the intentional body. In their accounts of *nervios*, on the other hand, these

Salvadoran women speak specifically of their perceived need to control themselves, to harness their anger and fear. In this regard, they are not unlike most other women worldwide who, relative to men, feel a disproportionate need to suppress their passions (Lutz 1990). The need for the domestication of emotion was evidenced, for example, in the common use *vapor* (vapor) as a metaphor. Since *vapor* is a term more normally employed within the domestic context of ideal cooking time (*al vapor*), the perceived need for the domestication of the raw (female) emotion is implicated. One woman described the experience of *calor* as the "stirring" of bodily memory throughout one's nervous system and blood.

While images of *vapor* and stirring evoke the notion of domestication, that of bodily memory evokes the engagement of affect associated with trauma, and incorporated in what Casey (1987) has called habitual body memory. Casey notes that traumatic body memory risks the fragmentation of the lived body such that it is

incapable of the type of continuous, spontaneous action undertaken by the intact body ("intact" precisely because of its habitualities, which serve to ensure efficacy and regularity). The fragmented body is inefficacious and irregular; indeed, its possibilities of free movement have become constricted precisely because of the trauma that has disrupted its spontaneous actions. (Ibid.: 155)

In the closely knit family atmosphere of Salvadorans, such trauma may be equally poignant if it occurs to immediate kin as if it occurs to the self. There is certainly ample narrative evidence for the importance of actual or near-death experiences of family members that may elicit the bodily sensation of *calor*. In sum, *calor* is the bodily channeling of emotions seen as emanating from without that must be thrown off, not only to regain a comfortable stance of being-in-the-world, but indeed to ensure one's very bodily integrity and survival.[15] As an existential phenomenon associated with *nervios*, *calor* is a vivid example of what Setha Low has described as an "embodied metaphor" (Low, Chapter 6 of this volume) of trauma that constitutes the habitual body memory of *la situación* for these women.

Medical and psychiatric diagnostic considerations

The clinical relevance of understanding *el calor* is all too evident in the following incident. While waiting for the resident to come into the hospital examining room, a patient was overcome by intense heat throughout her body. To relieve herself she took off her blouse and soaked it in cold water from the sink. When the resident entered the room and saw she was not only distressed but also half-nude, he apparently assumed she was "psychotic" and immediately transferred her to the local state psychiatric hospital, where she remained without the benefit of an interpreter for several days until her family discovered her whereabouts.

Clinical confusion over how to biomedically diagnose and treat *calor* was abundant in the women's narratives. Several of the Salvadoran women reported what they regard as common mis-diagnoses, including menopause or high blood pressure. The inadequacy of an explanation based on menopause is evident from the fact that *calor* is commonly experienced well before the onset of menopause (age range in the twenties and thirties) and the fact that men also experience *calor*.[16] Nevertheless, we cannot entirely rule out the interactive effects of menopausal symptoms among some women in the study. The overarching point is that the phenomenology of *calor* sensations varies substantially from symptoms of menopause, and that much if not all of the *calor* experiences described here are clearly not reducible to such explanations. In addition, *calor* (not unlike pain) is not clinically observable or measurable. The frustration surrounding clinical encounters was captured in one woman's comment that "I tell the doctors to look for something inside of me, but they tell me it's only what I feel."[17]

What are possible psychiatric renderings of *calor* experience? From the clinical perspective, these women were diagnosed as suffering from symptoms of one of more disorders including depression, panic attacks, generalized anxiety, and post-traumatic stress disorders. Major depression is especially common. In this connection, we note that the experience of "central heat" has been observed among depressed patients in societies such as Nigeria (Ifabumuyi 1981). The dynamics of this phenomenon in depression have yet to be fully appreciated in North American clinical practice. The *calor* component of the depression picture, however, may warrant identifying this phenomenological type of depression as "engaged" vs. "withdrawn," in so far as the women's subjective reports during the interview sessions did not, for us, have the "feel" of clinical depression.[18] Relevant to post-traumatic stress disorder, the occasional presence of dissociative states, numbness in the face or other body part (paresthesia and akisthesia), and "vigilant" startle response, all represent varying degrees of the fundamental human tendency to "flight or fight" in response to threat. The *calor* experience occurs both under conditions in which such a response is not only appropriate but essential, and under conditions that are possibly inappropriate or personally oppressive. This being said, it is important not to lose sight of the fact that *calor*, like *nervios* (Jenkins 1988) or depression (Kleinman and Good 1985), can in some circumstances be understood as normal, non-pathological experience. As one Salvadoran woman put it: "No pienso que es enfermedad" (I don't think it's illness).

The bodily reality of *el calor*: shared or culture-bound experience?

The foregoing discussion, as might be expected of any discussion that participates in a relatively new problematic, raises more questions than can be answered: how does the body "experience" itself, on the one hand, and "think" about itself, on the other? How does the body "select" certain things to experience or "think" about? Can any bodily experience or representation be considered apart from the affective context in which it occurs? Are all bodily experiences infused with "feeling" or "emotion"? Can bodily experience and representation be considered discrete sets of evidence or "symptoms" of "disorder" ("mental," "physical," "social," "emotional," or "cosmological")? Within the body of Latin American literature on *nervios*, how should *calor* be conceptualized? (1) as a subset of the general cultural categories of *nervios*? (2) as a discrete phenomenological experience of anger and fear in the context of gendered power inequities?

We would hope that our exposition of Salvadoran women's narratives of the experience of *calor* can serve in part as a basis for addressing these questions, and thus further the development of culture theory from the standpoint of embodiment. We conclude by identifying several problematic assumptions which must be scrutinized in the process of this development:
1. *The body is a* tabula rasa *upon which culture inscribes its codes.* This common, often implicit view of the relationship between culture and the body has tended to overlook the capacity of the body to generate cultural worlds of meaning in relation to being-in-the-world. In contrast, our standpoint assumes the body to be characterized by intentionality and agency.
2. *Personal experience of bodily-felt emotion is suspect on both theoretical and methodological grounds.* The notion that cultural discourse on emotion is superior to bodily-felt personal accounts runs the risk of contributing toward reductionistic accounts of affect; we note that studies of culture, emotion, and the body have typically not been carried out from the vantage point of experience.
3. *Because emotion is culturally constructed and not usefully conceived as psychobiologically universal, the biological (and hence the body) can reasonably be downgraded or excluded from our theoretical formulations.* Contrary to this assumption, we would argue that it is essential to understand the body as *both* cultural object and cultural subject. Cultural anthropology's flight from the biological ought not to have included the body.
4. *We can know only language and never experience.* We argue for the inherent inseparability of "raw" bodily experience at the immediate, sensate

level and its "cooked" linguistic, ethnopsychological representation, and further that this is equivalent to the inherent inseparability of being-in-the-world and our representations of experience.

5. *Culture theory is the primary starting and end point for studies of emotion.* On the contrary, cultural analysis that is conducted in the absence of sociopolitical considerations of power and interest is incomplete. In particular, as can be seen in our focus on women's lives to advance understandings of the body's emotional experience, it is essential to consider the highly gendered dimensions of these relations. The emerging agenda for studies of emotional processes and experience in any of an array of intentional worlds must explicitly incorporate political dimensions of such worlds large and small.

NOTES

1 My thoughts on emotion in relation to culture and "estrangement" originated in a somewhat separate context from consideration of Lutz's (1988) ethnopsychological analysis of emotion as "against" (or in contrast to) estrangement or disengagement in North American popular representations.

2 See Bateson (1958: 118) for a definition of ethos as the emotional environment of an entire culture and Jenkins (1991b) for an analysis of emotional atmospheres within families. Jenkins (1991a) conceives of political ethos as the culturally standardized organization of feeling and sentiment pertaining to the social domains of power and interest, and discusses its effects within particular family contexts among Salvadorans.

3 The authors wish to thank the editor, Thomas J. Csordas, and two anonymous reviewers for their comments. To the Salvadorans who participated in this study, we are much indebted.

4 The clinical setting is the teaching hospital of Harvard Medical School where Jenkins worked for three and a half years as a psychiatric anthropologist and Valiente worked as a clinical psychologist. Methods of the study included semi-structured conversational interviewing designed to obtain narratives of current and past life experience. With the exception of one informant, Jenkins conducted all ethnographic interviews and observations of home and community settings. The interview questionnaire was collaboratively constructed, with special attention to cultural and ethnopsychological issues by Jenkins and clinical and linguistic translation by Valiente. The majority of the patients in the present study were referred by Valiente.

Since the majority of the patients who utilize this clinic are women, and since women's refugee experience is distinctive from that of men's, a special focus on women was deemed appropriate both clinically and anthropologically. Depending upon the research participant, from two to fifteen interview or observational visits were completed. For a smaller set of participants who served as anthropological "key informants," and with whom Jenkins developed special rapport, visits were more numerous. The women in the study were between 20 and 62 years of age and primarily of peasant background, with little formal education. Most of the women were Catholic, monolingual Spanish-speakers.

5 Diagnostic data are in accord with Diagnostic and Statistical Manual III-R (American Psychiatric Association 1987) categories according to the Schizophrenia and Affective Disorders Schedule (SADS). Women in the study reported symptoms from a host of Axis I affective and anxiety disorders, including major depression, dysthymia, post-traumatic stress, generalized anxiety, somatization and panic disorders.

6 In a psychosocial study of motivation and achievement among Central American refugees who attend US high schools, Suarez-Orozco (1990) also reports common usage of the term *la situación*.

7 Regular, so-termed *"domestic"* violence and abuse are the bodily experience of many of the Salvadoran women refugees in the study. Indeed, some of them cited escape from abusive husbands and fathers as a principal reason for migrating from El Salvador.

8 See Jenkins (1988) for a discussion of the strategically broad nature and meaning of the category of *nervios*.

9 Special thanks to Jeff Jacobson and Maria-Jesus Vega for research assistance in the transcription, translation, and data organization of the interview material. Both have made contributions to the analyses developed here.

10 A *santero* is a religious leader or priest practicing within the religious tradition of *santéria*. See Harwood (1987) and Gonzales-Wippler (1989) for ethnographic accounts of *santéria*.

11 Kirmayer (personal communication) observes that the cross-cultural commonality of heat experiences has apparently been greatly underestimated. This may be due in part to Euro-American biases in the collection of basic ethnopsychological and ethnopsychiatric data.

12 For a critique of the concept of "belief" in anthropological theory, see B. Good (1994).

13 Adequate consideration of the extensive epistemological difficulties with cross-cultural application of the concept of somatization is beyond the scope of the present chapter; the most thorough treatment to date is to be found in the works of Kleinman (1986, 1988) and Kirmayer (1984).

14 The women who reported personal experiences of domestic violence would typically do so with shame. The cross-cultural commonality of violence by male kin within family settings is alarming (Levinson 1989).

15 Ethnopsychologically, this is likely to be also related to indigenous "hot–cold" theories common throughout Latin America.

16 Although our focus is on women's experience, ethnographic and clinical experience suggests that experiences of *calor* are not confined to women.

17 See Scarry (1985) for a discussion of the relationship between clinical gaze and the experience of pain.

18 This intersubjectively based observation is beyond the scope of this chapter. We briefly note here that, for these women, the interactive qualities of depression differ from those of other women diagnosed as suffering from clinical depression. This observation is based on ethnographer–informant encounters and clinical observations.

REFERENCES

Abu-Lughod, Lila (1986) *Veiled Sentiments: Honor and Poetry in a Bedouin Society*. Berkeley and Los Angeles: University of California Press.

Abu-Lughod, Lila and Catherine Lutz (1990) Introduction: Emotion, Discourse, and the Politics of Everyday Life. In Catherine A. Lutz and Lila Abu-Lughod, eds., *Language and the Politics of Emotion*, Cambridge: Cambridge University Press.

Bateson, Gregory (1958) *Naven*. Palo Alto, California: Stanford University Press (second edition).

Casey, Edward (1987) *Remembering. A Phenomenological Study*. Bloomington, IN: Indiana University Press.

Corradi, Juan, Patricia Fagen, and Manuel Garreton (1992) *Fear at the Edge: State Terror and Resistance in Latin America*. Berkeley and Los Angeles: University of California Press.

Csordas, Thomas J. (1993) Somatic modes of attention. *Cultural Anthropology* 8: 135–56.

(1994) *The Sacred Self: A Cultural Phenomenology of Charismatic Healing*. Berkeley: University of California Press.

Desjarlais, Robert (1992) *Body and Emotion: The Aesthetics of Illness and Healing in the Nepal Himalayas*. Philadelphia: University of Pennsylvania Press.

Doi, T. (1973) *The Anatomy of Dependency*, J. Bester, trans. New York: Harper and Row.

Farias, Pablo (1991) Emotional Distress and Its Socio-Political Correlates in Salvadoran Refugees: Analysis of a Clinical Sample. *Culture, Medicine and Psychiatry* 15: 167–92.

Fish, Joe and Cristina Sganga (1988) *El Salvador: Testament of Terror*. New York: Olive Branch Press.

Fernandez, James W. (1986) *Persuasions and Performances: The Play of Tropes in Culture*, Bloomington, IN: Indiana University Press.

Friedrich, Paul (1991) *Polytropy. In Metaphor: The Theory of Tropes in Anthropology*, James W. Fernandez, ed. Stanford, CA: Stanford University Press.

Frijda, Nico (1987) *The Emotions*. Cambridge: Cambridge University Press.

Gaines, Atwood and Paul Farmer (1986) Visible Saints: Social Cynosures and Dysphoria in the Mediterranean Tradition. *Culture, Medicine and Psychiatry* 10(4): 295–330.

Geertz, Clifford (1973) *The Interpretation of Cultures*. New York: Basic Books.

Good, Byron (1994) *Medicine, Rationality, and Experience: An Anthropological Perspective*. Cambridge: Cambridge University Press.

Good, Mary-Jo DelVecchio and Byron Good (1988) Ritual, the State and the Transformation of Emotional Discourse in Iranian Society. *Culture, Medicine and Psychiatry* 12: 43–63.

Good, Mary-Jo DelVecchio, Byron Good, and Michael Fischer (1988) Introduction: Discourse and the Study of Emotion, Illness and Healing. *Culture, Medicine and Psychiatry* 12, 1–8.

Gonzalez-Wippler (1989) *Santeria: The Religion. A Legacy of Faith, Rites, and Magic*. New York: Harmony Books.

Guarnaccia, Peter and Pablo Farias (1988) The Social Meaning of *Nervios*: A Case Study of a Central American Woman. *Social Science and Medicine* 26: 1231–3.

180   *Janis H. Jenkins and Martha Valiente*

Guarnaccia, Peter, Arthur Kleinman, and Byron J. Good (1990) A Critical Review of Epidemiological Studies of Puerto Rican Mental Health. *American Journal of Psychiatry* 147: 1440–56.

Haraway, Donna (1991) *Simians, Cyborgs, and Women: The Reinvention of Nature.* New York: Routledge.

Harwood, Alan (1987) *Rx: Spiritist as Needed: A Study of a Puerto Rican Community Mental Health Resource.* Ithaca: Cornell University Press.

Ifabumuyi, O. I. (1981) The Dynamics of Central Heat in Depression. *Psychopathologie Africaine* 17: 1–3, 127–33.

Jenkins, Janis H. (1988) Ethnopsychiatric Interpretations of Schizophrenic Illness: The problem of *nervios* within Mexican-American Families. *Culture, Medicine, and Psychiatry* 12(3), 303–31.

—→ (1991a) The State Constructions of Affect: Political Ethos and Mental Health among Salvadoran Refugees. *Culture, Medicine and Psychiatry* 15: 139–65.

(1991b) Anthropology, Expressed Emotion, and Schizophrenia. *Ethos* 19: 387–431.

—→ Kirmayer, Laurence (1984) Culture, Affect and Somatization, Parts 1 and 2. *Transcultural Psychiatry Research Review* 21(3): 159–88, (4): 237–62.

(1992) The Body's Insistence on Meaning: Metaphor as Presentation and Representation in Illness Experience. *Medical Anthropology Quarterly* 6(4): 323–46.

Kleinman, Arthur (1986) *Social Origins of Distress and Disease. Depression, Neurasthenia, and Pain in Modern China.* Yale University Press.

(1988) *Rethinking Psychiatry.* New York: Free Press.

(1992) Pain and Resistance: The Delegitimation and Relegitimation of Local Worlds. In M. J. DelVecchio Good, P. Brodwin, B. Good, and A. Kleinman, eds., *Pain as Human Experience: An Anthropological Perspective.* Berkeley: University of California Press.

Kleinman, Arthur and Byron Good, eds. (1985) *Culture and Depression: Studies in the Anthropology and Cross-Cultural Psychiatry of Affect and Disorder.* Berkeley: University of California Press.

Levinson, David (1989) *Family Violence in Cross-Cultural Perspective.* Newbury Park, CA: Sage Publications.

Levy, Robert (1984) Emotion, Knowing, and Culture. In Richard A. Shweder and R. A. LeVine, eds. *Culture Theory: Essays on Mind, Self, and Emotion.* Cambridge: Cambridge University Press.

—→ Low, Setha (1985) Culturally Interpreted Symptoms or Culture-Bound Symdromes: A Cross-Cultural Review of Nerves. *Social Science and Medicine* 221(2): 187–96.

(1988) The Diagnosis and Treatment of *Nervios* in Costa Rica. In Margaret Lock and Deborah Gordon, eds. *Biomedicine Examined.* Dordrecht: Kluwer Academic Publishers.

Lutz, Catherine (1985) Ethnopsychology Compared to What? Explaining behavior and Consciousness among the Ifaluk. In Geoffrey M. White and John Kirkpatrick, eds., *Person, Self, and Experience: Exploring Pacific Ethnopsychologies.* Berkeley and Los Angeles: University of California Press.

(1988) *Unnatural Emotions: Everyday Sentiments on a Micronesian Atoll and Their Challenge to Western Theory.* Chicago: University of Chicago.

(1990) Engendered Emotion: Gender, Power and the Rhetoric of Emotional Control in American Discourse. In Catherine A. Lutz and Lila Abu-Lughod, eds., *Language and the Politics of Emotion*. Cambridge: Cambridge University Press.

Lutz, Catherine and Lila Abu-Lughod, eds., (1990) *Language and the Politics of Emotion*. Cambridge: Cambridge University Press.

Lutz, Catherine and Geoffrey White (1986) The Anthropology of Emotions. *Annual Review of Anthropology* 15: 405–36.

Markus, Hazel and Shinobu Kitayama (1994) *Emotion and Culture: Multidisciplinary Perspectives*. American Psychological Association Press.

Martin-Baro, Ignacio (1990) De la guerra sucia a la guerra psicologica. *Revista de Psicologia de El Salvador* 35: 109–22.

Matthews, Holly (1992) The Directive Force of Morality Tales in a Mexican Community. In R. D'Andrade and C. Strauss, eds., *Human Motives and Cultural Models*. Cambridge: Cambridge University Press.

Myers, Fred (1986) *Pintupi Country, Pintupi Self: Sentiment, Place, and Politics among Western Desert Aborigines*. Washington: Smithsonian Institution Press.

Ochs, Elinor and Bambi Schieffelin (1989) Language Has a Heart. *Text* 9: 7–25.

Ortner, Sherry B. (1976) Is Female to Male as Nature is to Culture? In Michelle Rosaldo and Louise Lamphere, eds., *Woman, Culture, and Society*. Stanford: Stanford University Press.

Ots, Thomas (1990) The Angry Liver, the Anxious Heart and the Melancholy Spleen. *Culture, Medicine and Psychiatry*, 14: 21–58.

Rosaldo, Michelle (1980) *Knowledge and Passion: Ilongot Notions of Self and Social Life*. Cambridge: Cambridge University Press.

(1984) Towards an Anthropology of Self and Feeling. In R. A. Shweder and R. A. LeVine, eds., *Culture Theory: Essays on Mind, Self and Emotion*. Cambridge: Cambridge University Press.

Rosaldo, Michelle and Louise Lamphere, eds., (1974) *Woman, Culture, and Society*. Stanford: Stanford University Press.

Roseman, Marina (1991) *Healing Sounds from the Malaysian Rainforest: Temiar Music and Medicine*. Berkeley and Los Angeles: University of California Press.

Scarry, Elaine (1985) *The Body in Pain: The Making and Unmaking of the World*. New York: Oxford University Press.

Scheper-Hughes, Nancy (1992) *Death without Weeping: The Violence of Everyday life in Brazil*. Berkeley and Los Angeles: University of California Press.

Scheper-Hughes, Nancy and Margaret Lock (1987) The Mindful Body: A Prolegomenon to Future Work in Medical Anthropology. *Medical Anthropology Quarterly* 1(1): 6–41.

Schieffelin, Edward (1976) *The Sorrow of the Lonely and the Burning of the Dancers*. New York: St. Martin's Press.

Shweder, Richard A. (1990) Cultural Psychology: What Is It? In James Stigler, Richard Shweder, and Gilbert Herdt, eds. *Cultural Psychology: Essays on Comparative Human Development*. Cambridge: Cambridge University Press.

Shweder, Richard A. and Robert A. LeVine, eds. (1984) *Culture Theory: Essays on Mind, Self, and Emotion*. Cambridge: Cambridge University Press.

Suarez-Orozco, Marcelo (1990) Speaking of the Unspeakable: Towards a Psychosocial Understanding of Responses to Terror. *Ethos* 18: 353–83.

Swartz, Leslie (1991) The Politics of Black Patients' Identity: Ward-Rounds on the

'Black Side' of a South African Psychiatric Hospital. *Culture, Medicine and Psychiatry* 15(2): 217–44.

White, Geoffrey M. and John Kirkpatrick, eds. (1985) *Person, Self, and Experience: Exploring Pacific Ethnopsychologies.* Berkeley and Los Angeles: University of California Press.

Wikan, Uni (1990) *Managing Turbulent Hearts: A Balinese Formula for Living.* Chicago: University of Chicago Press.

The embodiment of symbols and the
acculturation of the anthropologist

Carol Laderman

Symbolic systems exist on a number of interpenetrating levels, ranging from
the most abstract to the most concrete. All contribute to the maintenance of
the system. The most abstract level reflects and gives coherence to a world
view. Its language and theory are logically consistent throughout a number of
conceptual domains. A middle level of abstraction provides metaphors for
reasoning in the mundane world. The most concrete level is direct sensory
experience which moors the system at strategic points to empirical reality
and acts as a structural support to the symbolic edifice. Its "proof" of the
system's validity is most persuasive when it occurs in the context of illness or
other situations perceived as dangerous, and when effect appears to follow
closely after cause. The symbolic structure makes experience meaningful,
but without an input from sensory reality the edifice would crumble.

Four salient aspects of the embodiment of Malay symbolism were
explained to me by informants, and became, as well, part of my personal
experience. According to Malay theory, we are all born with four bodily
humors, varieties of Inner Winds (*angin*) that determine personality, and the
spirit of life (*semangat*). Some people have, as well, a tendency towards dis-
harmony that only manifests after eating certain foods (*bisa*). These components of the self must be protected from the harm that occurs when,
through loss of inner balance, they are depleted, become overabundant, or
are seriously skewed. Balance is the key to well-being in the individual, the
community, and the universe – the harmonious interaction of the microcosm
and the macrocosm, the person and society, the internal and the external.

The first section of this chapter will discuss the embodiment of Malay
symbolism. The second section will discuss the acculturation of this
embodiment as it was experienced by the anthropologist.

I

The humors

Humoral reasoning, based upon belief in the universality of four basic
elements – earth, air, fire, and water – and their manifestations in the

cosmos, the body politic, and the human person, pervades contemporary Malay thought. Its metaphors center around notions of heat and coolness which, in a humoral system, refer to intrinsic qualities rather than merely thermal temperatures, i.e., alcohol is always "very hot" humorally speaking, even when served on ice; steaming hot squash is still humorally "cold."

Malay identification of heat with destruction and coolness with well-being, a mirror image of ancient Greek humoral doctrine which equated heat with vitality and coolness with ill-health, is reflected in daily language. *Sejuk* ("coolness") can be used as a synonym for "healthful, energetic, and pleasant." *Menyejukkan* ("to make cool") can mean "to calm, revive, repair, amuse." A person whose liver (the Malay seat of emotions) is cool is tranquil and carefree. In contrast, *panas* ("heat") can be a synonym for unlucky, ominous, disastrous. Those with *panas rezeki* ("hot livelihood") are poor unfortunates. An ill-tempered person is described as a "glowing ember," a person quick to anger has a "hot liver" or "hot blood." "To cause the liver to become hot" means to instill hatred in one's breast. Black magic is called "the hot science" because it carries out a mission of hatred (Awang Sudjai Hairul and Yusoff Khan 1977; Iskandar 1970; Wilkinson 1959).

The power of the Malay ruler to keep or restore peace and harmony within his kingdom rests, metaphorically, in his ability to provide the coolness that balances destructive heat threatening the body politic. If his rule is harmonious, the heat of war will not destroy the nation, the heat of anger will not cause internal dissension, and the heat of nature will not destroy the crops. Within his own person, the successful ruler embodies this coolness, made explicit by such expressions as *"Perentah-nya sejuk,"* usually translated as "His reign was full of benign influences," but literally meaning "His reign was cool" (Zainal-Abidin 1947: 43).

In order to maintain harmony, the invisible disembodied spirits must not be allowed to encroach upon the human domain. Historically, the ruler was considered to be not only the defender of Islam but also a harborer of friendly familiar spirits and controller of spirits whose heat puts the kingdom at risk, spirits incomplete in the universal scheme of things, lacking the earthly and watery elements of which the human body is made. Control of these spirits, on the national level, was the duty of the state shaman, usually a brother of the sultan. Commoners also contributed to the maintenance of national harmony by their daily recitation of the "blessed cooling prayers" of Islamic obligation (ibid.: 43).

On a personal level, Malays depend upon their traditional healers (*bomoh*) to deal with illnesses brought on by spirit attacks, as well as those with more mundane causes. Spirits most often afflict human victims by blowing superheated breath on the victim's back, upsetting his humoral balance. The *bomoh* has several methods of increasing cold and wet (earthy and watery) elements of the patient's body. Spells are recited with the patient's back to the healer. At strategic times during his recitation the *bomoh* blows on the patient's back. His breath, cooled by the incantation, counteracts the spirits' hot breath. Illness due to spirit attacks is rarer than the fevers, respiratory ailments, and digestive upsets which are believed to result from a humoral imbalance. These might occur because of improper management of diet, work, sleep, because one's body has not adjusted to changes in the weather, or other problems of daily life. Malays generally agree on criteria for classifying ordinary sicknesses, which interpret sensory evidence on the basis of humoral reasoning. Humoral concerns are very salient in contemporary life. Within days after I arrived at my research site I was warned that eating *durian* fruit together with hospital-type medicines was dangerous since their combined heat might cause madness or even death.

There are several rules of thumb Malays use to decide which foods are humorally hot or cold, or belong to a third category, *sederhana*, neutral or the proper mean (see Laderman 1983). Hot foods include fats, oils, animal flesh, spices, salty and bitter foods, and alcohol. Most cold foods are juicy, slimy, sour, or astringent. The ultimate test is the effect on the body, but while a majority claimed they could feel some heating or cooling effects of food, not everyone is sensitive enough to perceive subtle differences. Appealing to sensory perceptions is further complicated by the dynamic nature of the humoral system whose model incorporates variability. Individuals differ physically and temperamentally; all change as they pass through stages of life, seasons of the year, and hours of the day. Thus, a person who tends toward the hot polarity would classify as neutral a food considered "cold" by most people, since his own body does not feel its cooling effects as strongly as others might.

Empirical observation reinforces the humoral world view in the eyes of its adherents. Western science, while rejecting humoral reasoning, has provided some support for these "folk" observations. Most "cold" foods contain large amounts of indigestible roughage, so one would expect they would irritate some people's digestive tracts while leaving others unscathed. Eating only "cold" foods is likely to lead to weakness, since their composition does not provide complete proteins, concentrated food energy, or fat-soluble vitamins. Fats and oils, humorally "hot," make people feel satiated since they provide nine calories per gram, compared with four per gram for carbohydrates. Animal proteins, also "hot," contain all essential amino acids necessary for the maintenance of health, compared with the incomplete amino acids provided by "cold" fruits and vegetables. After ingestion of proteins, the body's heat output is increased by 20 to 30 percent over intake, compared with 5 percent for ingestion of carbohydrates (Burton 1965: 25). Salt, and foods high in sodium, have been shown to aggravate hypertension in many susceptible sufferers. In an interesting experiment (Ramanamurthy 1969), four individuals were put on "hot" diets, according to Indian humoral conceptions, for ten days, and then placed on "cold" diets for the same length of time. Subjective feelings reported during the "hot" diet were burning eyes, burning urination, and a general feeling of warmth in the body. Analysis of urine and feces showed that those on the "hot" diet displayed higher urine acidity and sulfur excretion and lower retention of nitrogen.

Bisa

Less pervasive than the humoral system but even more important within its narrow range is the concept of *bisa*. Like the humoral system, it is strengthened by sensory phenomena that, through their force, carry along the belief system as a whole. The dictionary definition of *bisa* is "blood-poison, anything that gives a septic wound, venomous. Of the stings of hornets, scorpions and centipedes; the bites of snakes; the poison used on darts; the septic nature of bites of tigers and crocodiles, and of wounds from krises of laminated steel" (Wilkinson 1959: I: 145). This definition leaves out the core meaning of the term and neglects its most common usage. Malays refer to power, whether used for good or evil, as *bisa*. The words of the *bomoh*, neutral when he is not speaking in his professional capacity, become *bisa* when he chants an incantation. The words of the sultan are always *bisa* to his subjects. No moral judgment is implied: the *bomoh*'s *bisa* words are usually used for their curative power, although he may also use them to harm; the ruler's *bisa* speech is for the well-being of the nation, although it may prove painful to some of his subjects.

The most frequent everyday use of *bisa* in rural Malay society is in connection with foods and food avoidance. It has been treated by social and medical scientists as symbolism about impure food "succinctly expressive of the patient's dilemma as a social being" (Provencher 1971: 188–9), or as superstitious beliefs detrimental to health (Wilson 1970, 1971; Mills 1958: 141). Rural Malays, on the other hand, believe that avoiding *bisa* foods has a beneficial effect on health. Knowledge of which foods are *bisa* is common; people act upon it during illnesses and in other vulnerable conditions, particularly the postpartum period. Beliefs about *bisa* foods accord well with what is known about allergies. Allergic reactions occur only after the patient has been sensitized to a particular allergen, and may be lifelong or transient. Malays believe that *bisa* foods exacerbate preexisting disharmonies in individuals who may have been unaware of this condition. Illness may hide unknown within a person's body, emerging only after *bisa* food is eaten. It would be a mistake, however, to assume that *bisa* and allergen are synonymous. Many of my Malay neighbors developed red itchy welts after eating albacore or Spanish mackerel and immediately classified these fish as *bisa*. Their welts may have been due to scombroid poisoning. Fish normally contain a chemical constituent called histadine, found in varying amounts in different species. When histadine is acted upon by bacteria it changes into saurine, a histamine-like substance which can cause an illness resembling severe allergy. Scombroid fish are especially prone to become toxic when left to stand in the sun, or even room temperature in the tropics (Halstead 1959: 112; 1967), but do not cause illness when fresh. It is significant that *bomoh* advise patients not to eat stale fish. Their dietary recommendations may be grounded in the reality, as well as the symbolism, of impure food.

The concept of *bisa* is most complete during postpartum and postcircumcision periods. Until the wound of circumcision is healed, a boy adheres to the same diet prescribed for the postparturient woman. They avoid "cold" fruits and vegetables, fried foods, mollusks, crustaceans, and an extensive list of fish. The midwife refers to these fish in particular when she cautions patients against eating *bisa*. The fact that some women in the puerperium have eaten these fish without experiencing problems does not affect a general belief in the validity of the system. Malays reason that some people are so strong, or lucky, they can get away with dangerous actions. The epidemiology of scombroid, ciguatera, and other fish poisoning offers ideal support for these beliefs. Scombroid poisoning does not invariably occur when one eats scombroid fish, and does not affect every victim with equal severity; ciguatera is not invariable after eating implicated species. The incidence and severity of these illnesses depend on the potential victim's prior state of health as well as the level of toxicity in the individual fish and the amount eaten. Although more than 300 different species have been incriminated in ciguatera, its occurrence is unpredictable and therefore exceedingly difficult to control (Halstead 1959: 117). An occasional health problem associated with specimens of a particular species can be enough to place the entire species, and possibly related species, in the *bisa* category. Within any species, only some individuals are toxic. Of two caught simultaneously in the same place, one could be eaten without ill effects and the other could produce agonizing symptoms (Gordon 1977: 223). The variable nature of this input from sensory perception supports rather than vitiates Malay belief regarding the wisdom of avoiding *bisa* fish during vulnerable periods.

Semangat

Maintenance of good health requires more than just a proper diet and sensible life style. Following a world view that harmony and balance support the health of the cosmos, the body politic, and the human body, one must not only guard against tipping oneself too far toward a humoral polarity; one must also protect one's self against depletion or overabundance of other component parts. The person, in the Malay view, is composed of more than a thinking mind, a body that decays after death, and a soul that lives on in Heaven or in Hell. Some of its parts are common to all of creation, some shared with the animal kingdom, others restricted to humans, and others particular to individuals.

Human beings, like all God's creatures, must inhale the Breath of Life at birth. This *nyawa*, containing the elements of air and fire, animates the watery, earthy body; without it the body must die. It drives the blood in its course; its effects are felt within the body and its presence is obvious to observers when it emerges as breath, just as a breeze, itself invisible, is signaled by the rustling of leaves and the feeling it produces as it blows on the skin.

Semangat (Spirit of Life) is not limited to animals. It permeates the universe, dwelling in man, beast, plant, and rock. The universe teems with life: the life of a fire is swift; a rock's life is slow, long, and dreamlike. *Semangat* strengthens its dwelling place, whether the human body or a stalk of rice, and maintains health. It is extremely sensitive, however, and can be depleted: it may even flee, startled, from its receptacle. The vulnerability of *semangat* governs the conduct of the traditional Malay rice harvest. Modern methods may be more efficient, but they are not calculated to spare the feelings of the Rice Spirit (*semangat padi*). To a traditional Malay, the field of rice is like a pregnant woman, and the harvest is equivalent to the birth of a child. It is inaugurated by the taking of the Rice Baby, a stalk of rice swaddled like a human child after being cut from its plant with a small, curved blade concealed in the hand, so as not to frighten the Rice Spirit by its brutal appearance. The harvested rice crop is stored in a special bin with a coconut, coconut oil, limes, *beluru* root shampoo, bananas, sugar cane, water, and a comb, all for the use of the Rice Spirit, personified as a timid young woman (see also Firth 1974: 192–5). Since she is easily frightened, the rice must be brought back to the storeroom and left there in silence for three days.

Semangat in humans is similarly timid and must be protected. It can be summoned by spells such as thwarted lovers use to regain their beloved, or called by a *bomoh*, using the same sound Malays use to call their chickens (kurrr). The timidity of *semangat* makes it prone to leap and fly at the approach of a frightening object or unexpected noise. The effect of the startle reaction is most serious in people who are already in a vulnerable condition, such as pregnant women whose fear may be communicated to their unborn children, resulting in infantile abnormalities. *Semangat* loss is felt as loss of energy, loss of confidence, a weakness in body and mind.

Although most people live secure within the "gates" of their individuality, some people's boundaries are riddled with tiny openings, more like a permeable membrane than a wall. At times, such people find their thoughts becoming confused, their actions less than voluntary. The extraordinary permeability of some individuals was offered to me by many Malays as an explanation of *latah*, a condition in which being startled by a loud noise or an unexpected event triggers a spate of obscene language or imitative behavior. This startle reaction in turn increases the permeability of the membrane, allowing the thoughts of others to mix with those of the *latah* victim's own and to govern his actions.

Angin

Inner Winds (*angin*) determining the child's personality, drives and talents are already present at birth. Their presence, type and quality can be deduced from the behavior of their possessor, but they are palpable neither to observers nor to their owner, except in trance, when they are felt as high winds blowing within the possessor's breast.

Angin is a word with multiple meanings, many of which are connected with notions of sickness and treatment. It can refer to the wind that blows through the trees, a wind that may carry dirt and disease. A strong cold wind can make you sick if it chills your body, upsetting your humoral balance and causing upper respiratory symptoms and pains in the joints. *Angin* in the stomach is produced spontaneously when a person overeats, making his belly swell and producing heartburn and nausea. Diseases that are not suspected of being a result of spirit attacks but are not readily diagnosed are often called wind sickness, meaning essentially, "I don't know what it is."

Another meaning of *angin*, capricious desires, is closer to its core meaning within the shaman's seance. If these whims are not indulged, the whimsical one may feel a sadness known as wind within the liver. The concept of *angin*, however, goes far beyond the whimsical, and the thwarting of these Inner Winds can result in consequences more serious than sadness.

The Inner Winds, as understood by east coast Malays, are close to Western concepts of temperament, both in the medieval sense of the four temperaments and as artistic temperament. They represent the airy part of the four universal elements, as shown in the shaman's divination. He counts out grains of popped rice into three piles. If the count in any pile ends on fire, the diagnosis points to a hot illness and may implicate the hot breath of spirits or the hot anger of humans as a cause of the patient's suffering; earth implies a cold illness, possibly caused by spirits of the earth; water points to a phlegmy condition, possibly contracted near the river or ocean. Wind in the divination refers to problems with the Inner Winds. All people are born with *angin*, the traits, talents and desires representing their ancestors' heritage, but some have more, or stronger, *angin* than the common run. If they are able to express it, they can lead untroubled lives and be respected for their strong and gifted characters. If they cannot, their *angin* is trapped inside them where it accumulates and produces *sakit berangin*, sickness due to blockage of the Inner Winds. Euro-Americans recognize this problem in artists and writers whose creativity is blocked, or whose art is insufficiently appreciated, and would not find it difficult to understand why Malays say that musicians, actors and puppeteers are attracted to their professions because of *angin*, and could not succeed without it. Malays do not recognize a split between the arts, the sciences, and sports; what we call artistic temperament refers to a wider range of behavior among east coast Malays than it does in Euro-American culture. Most of the healers are also performers in the Malay opera (*Mak Yong*) or shadow play, or are masters of the art of self-defense. Healers of all types must possess *angin* specific to their calling and suffer when their talents are ignored. A *bomoh* whose patients have forsaken him, a midwife with limited mobility, even a masseur without a steady call on his services can develop *sakit berangin*. Malays take the concept yet further. Everyone who hopes to be successful in any of the specialized roles that Malay village society provides must not only study diligently but also have the *angin* specific to that role. No amount of study can substitute for *angin*. Those whose *angins* are not appropriate to their roles may be particularly at risk of *sakit berangin*, such as a man who has inherited the *angin* of a midwife but has no opportunity to assist a woman in childbirth.

The meaning of *angin*, and the problems it may entail, extend beyond professional temperament to the basic personality. The majority of conditions treated by *Main Peteri*, the shamanic performance which cures by means of singing, dancing, trance, and dramatic characterizations of spiritual forces (Laderman 1991), are *sakit berangin*, and the most prevalent variety is due to thwarting of the personality type known as *Angin Dewa Muda* (The Wind of the Young Demigod), whose needs are those of royalty: fine clothing, delicate food, aromatic perfumes, comfortable living, and the love and respect of kin, friends and neighbors. Many people have inherited *Angin Dewa Muda* but few can satisfy its demands. Such people need to be pampered and admired, provided with life's luxuries and reassured often of their worth. Malay village society, where neither material goods nor overt expressions of affection and admiration are in plentiful supply, is a difficult setting for this personality. It is not enough to say that the expression of strong emotions is frowned upon in rural Malaysia. Most people deny that they have any. Women suffer through prolonged labors without raising their voices, and mourners at funerals remain dry-eyed. Married couples normally exhibit no signs of connubial affection or public aggression.

I would never have known that my next-door neighbor was almost strangled to death one night by her mistakenly jealous husband had she not shown me his finger marks on her throat. No tell-tale sound had broken the peace of the night. Her husband is a classic example of *Angin Hala*, the Wind of the Weretiger, which makes one quick to anger and heedless of its consequences. *Angin Hala* is difficult to express unless its possessor is a fighter or occupies a social position that allows him to vent his aggression without fear of retaliation. Those with tigerish personalities may prove dangerous to others if they express their *angin*, and dangerous to themselves if they do not.

Angin Dewa Penchil, an archetype from the Malay opera, is the heritage of nobility dissatisfied with their lives and homes. They wander in foreign parts and dress and behave like people of lower status, confusing the rules and etiquette of proper society. This kind of inappropriate behavior, whether it concerns the denial of special prerogatives by aristocrats or, even more worrisome, the dangerous presumptions of equality by those of humble origins, has been discouraged by Malay law and custom. In the past it was a punishable offense for commoners to use royal language when referring to themselves. Aside from legal sanctions against transgressing the prerogatives of royalty, commoners who dared break the social barrier or disobey royal commands might expect to fall ill with an incurable skin disease, or suffer in other respects from the curse that accompanies the flouting of royal power (Mohd. Taib Osman 1976).

People may be heir to one or several types of *angin*. Their strength can range from a mild breeze to gale force. These Winds, freely blowing or sublimated in ways that satisfy both possessor and society, keep the individual healthy and enrich his community. A person with *Angin Dewa Muda* may try to earn the love and admiration he so desperately needs. A man with *Angin Hala* may cover himself with glory on the playing field or battlefield. Powerful *angin*, ignored or repressed, will make its effects felt in the mind and body. Its symptoms include backaches, headaches, digestive problems, dizziness, asthma, depression, anxiety; in short, a wide range of what we call psychosomatic and affective disorders. Asthma in particular represents a graphic example of repressed *angin* – Wind locked within, choking its possessor.

The Inner Winds of *Main Peteri* patients who have been diagnosed as suffering from *sakit berangin* must be allowed to express themselves, released from the confines of their corporeal prison, enabling the sufferer's mind and body to return to a healthy balance. The band strikes up appropriate music as the shaman retells the story of the *angin*'s archetype. When the correct musical or literary cue is reached, the patient achieves trance, aided as well by the percussive sounds of music and the rhythmic beating of the shaman's hands on the floor near the patient's body. The essential differences between the patient's trance and that of the shaman are (1) the patient's trance does not make him or her a conduit for spirits, but, rather, puts patients in touch with their inner being; and (2) shamans control the alterations of their own consciousness, while the patient's consciousness is controlled by the shaman.

In shamanic ceremonies of other cultures, whose primary aim is to remove demonic influence from human sufferers, the patient's trance is "the peak moment [at which] the object of the demonic enters into direct communion with the subject" (Kapferer 1983: 195). For Malay patients, trance does not occur during exorcistic parts of the *Main Peteri*. The communication is not with the demonic, but with their own inner nature. While in trance, if a patient has *angin* for *silat* (the Malay art of self-defense), he will rise and perform its stylized moves and stances; if she has *angin* for *Mak Yong*, she will dance with the grace of a princess; *Angin Hala* can cause a trancing patient to roar and leap like a tiger. Patients are encouraged to act out the repressed portions of their personalities until their hearts are content and their *angin* refreshed. Coming out of trance, into the awareness of an enthusiastic, approving audience, the patient experiences a wonderful feeling of relaxation and satisfaction. Headaches and backaches have disappeared, and asthma sufferers find they can once again breathe freely.

II

Embodiment and the acculturation of the anthropologist

Anthropologists use their own bodies and minds as primary tools for the investigation of cultures. They participate as deeply as possible in the lives of those they study, at the same time maintaining sufficient distance to observe the workings of culture. They become insider–outsiders, observing with cool eyes and participating with a warm heart. Anthropologists who have achieved rapport find that they have become acculturated to some degree to their informants' beliefs and behaviors. Signs of this process are changes in bodily movements and postures, as when I assumed the squatting position used by Malay women while preparing foods for cooking. I hardly realized I was squatting rather than sitting at a table to perform this action as I had done throughout my life, until my husband laughingly remarked on it, calling me his "Malay wife." Another sign is change in habits of the mind: many of the characters in my dreams, including myself, spoke in Malay, and the symbols of my night theater were symbols of Malay, rather than Freudian, thought.

My neighbors were so convinced of the empirical reality of humoral doctrine that, when I sent off foods to be analyzed by the Institute for Medical Research, they assumed that they would not only be tested for nutrient content but scientific procedures would finally put to rest disagreements as to their humoral qualities. This insistence on the reality of humoral reasoning led me to observe my own physiological reactions to foods. I found "cold" foods refreshing, especially in the hot weather. Unlike some of my neighbors, I was able to eat them in large quantities at any time of the day or year without ill effects. According to the Malays, this was due to my body's unusually high intrinsic heat. This became obvious to them when they noticed how I thrived during the monsoon season, whose cold temperatures left them shivering under layers of sweaters. Even compared to those who live in temperate zones, my body is unusually warm in sickness and in health: my family closes windows I have thrown open in the winter, and my doctor no longer panics when my fever rises to 104 degrees. In the case of humorally hot foods, I found my sensitivity only extended to those considered extremely hot. Eating large quantities of *durian* fruit produced a feverish sensation, a feeling of great satiety, and a tendency toward diarrhea (which Malays attribute to heat softening feces). *Durian* had the same effects on my neighbors, but during its season we all stuffed ourselves on its delicious flesh, ignoring any attempt at a balanced diet and oblivious of the consequences. Not being pregnant during my time in Malaysia, I was not considered vulnerable to the effects of most *bisa* fish. As a lifelong hater of fish, eating only small amounts out of necessity, I avoided exposure to the risks of eating large quantities of scombroid fish.

The most striking examples of my acculturation came from my experience of the embodiment of components of the self as understood by Malays. As part of a shaman's entourage, and later his "daughter," I spent much of my time observing shamanic rituals and interviewing healers and their patients. I was assured that the ceremonies I observed were just the tip of the iceberg: many of its dramas occurred unseen by observers but felt by participants. When I asked people how it felt to be in trance, they couldn't, or wouldn't, answer. They told me that the only way I could know was to experience it. I avoided the issue, feeling uncomfortable with the lack of control that trance implied. Well into the second year of my research, while attending a *Main Peteri* as part of the shaman's entourage, he motioned to me to sit down on the mat recently vacated by his patient. I thought he wanted to do a short ritual to release me from the dangers inherent in witnessing women give birth, which he had often performed for my benefit. Instead, he proceeded to recite the story of *Dewa Muda* (which he had deduced was my primary Inner Wind), accompanied by the band and his own rhythmic pounding on the floor. My trust in him was strong enough to allay my fears, and I allowed my consciousness to shift into an altered state. At the height of my trance, I felt the Wind blowing inside me with the force of a hurricane. When I later described my feelings in trance, people assured me it was a common experience. They also wondered at my surprise. One woman remarked, "Well, why did you think we call them Winds?"

My experience of *semangat* as a component of my self came as a result of its loss. My second period of research on Malaysia's east coast coincided with an epidemic of dengue hemorrhagic fever, carried by day-flying *Aedes* mosquitoes. I must have been incubating it during the last days of my research. Shortly after my arrival in New York City, my temperature rose to 107 degrees. I was admitted to the hospital, placed on an ice mattress, and covered with ice cubes. My doctor ordered several tests, including a bone marrow tap. Since this was a teaching hospital, it proved to be a great opportunity for a novice to practice his technique, removing a marrow sample from my pelvis. It took him four tries before success. The pain was exquisite. As waves of agony coursed through my body, part of me seemed to flee from its normal dwelling place. I had never responded to pain in quite those terms, although I certainly had experienced more intense and longer-lasting suffering several years before, when I was hospitalized with a case of severe pelvic inflammatory disease.

After the doctor had obtained his marrow sample, I felt as though some essential component had left its fleshy confines and was hovering high above my head. I felt depleted, lacking in energy and a prey to emotions. My nerves literally seemed to be on the surface of my body; my skin appeared to offer no protection. I left the hospital two weeks later, but feelings of depletion and vulnerability remained with me months later. I seemed to have lost part of my self, the part that Malays call *semangat*, the spirit of life.

A friend, then the editor of a popular health magazine, had recently written an article about her experiences in a sensory deprivation flotation tank. She hoped the relaxation it usually produces would be good for me, and arranged for me to have a session.

Before entering the tank, one must wash away the oil that normally covers the skin, thus removing any barrier between the body and the water of the tank. The tank is approximately seven feet long, four feet wide, and four feet high, half-filled with water that has been super-saturated with epsom salts until, like the Dead Sea, it can support a barely submerged body. After one enters the tank the hood is closed, leaving one in total darkness and silence. All one can hear are the whispers of one's breath, the beating of one's heart and the sound of blood coursing through the veins. There is no sense of time, no sense of space. I had lain there for an unknown length of time when I felt my body begin to undulate up and down, then slowly spin in a complete circle, a physical impossibility since my body is much too long for the width of the tank. Finally the spinning stopped, and it was then I saw the birds. First, there were flocks of swifts which flitted about, gathered together and coalesced into one brilliantly colored cockatoo. The cockatoo soon changed to a giant Garuda bird, the mount of Vishnu, but instead of the god, it was I who was riding on its back at the same time that my body was still resting on the water. The bird and its mount flew from a great distance closer and closer to my body until we merged and became one. A feeling of wholeness and joy infused me; I knew my *semangat* had returned. After leaving the tank and showering once again to remove the salt, I experienced another change: my nerves were no longer exposed; they felt as though they had been covered by protective velvet.

That night I mused over my experience. I had hoped to benefit from the relaxing time my friend had described as usual for a float in a sensory deprivation tank, but I was completely surprised by the arrival of the birds. What did it mean? Why had I envisioned birds, and why those particular birds?

Semangat is often represented in Malay thought by a bird. Swifts, as common in Malaysia as sparrows in New York City, seem to be forever busy, always darting here and there. I was seen by my Malay neighbors as someone who rarely rested, never hesitating to stay awake all night to help deliver a baby or attend a healing ceremony, always willing to stop what I was doing to drive those in need to the hospital. The swifts seemed to me to represent that very visible part of my personality. The cockatoo symbolized another facet of my self, the *Dewa Muda* that loves beautiful clothes and jewelry, and revels in the admiration of others. The Garuda bird and its rider, Vishnu, belong to the Malays' religious past. For a thousand years before their conversion to Islam, Malays were Hindus. The high gods of Hinduism still exist in Malay mythology, often appearing in healing ceremonies as beneficent entities who speak through the voice of the shaman. Vishnu the Preserver was a highly appropriate symbol for the return of the spirit of life to an anthropologist who had so recently been absorbed in the study of Malay healing performances.

The symbol and the flesh

Noting the physical effects of eating, feeling the Inner Winds blowing in their bosoms during trance, and experiencing the loss and return of *semangat* would be incontrovertible proof to most east coast Malays of the validity of humoral reasoning and the existence of *angin* and *semangat* as essential parts of the self. I do not argue that Malays have discovered *the* truth about humanity and the workings of the universe. Experiences similar to mine and those of my Malay informants, had they been felt by people acculturated to other symbolic systems, would have been interpreted differently or assumed a different shape. Biomedical science, for instance, would offer a non-humoral explanation for the heat I felt after gorging on *durian*, pointing to the physiological effects of ingesting large quantities of fat. Explanation of my feelings in trance might be attributed to suggestion, catharsis, excitation of the nervous system, or production of endogenous chemicals.

The content of visions varies with the content of waking thoughts and assumptions about reality. South American Indians under the influence of Datura see jaguars and serpents, while medieval European believers in witchcraft, after taking the same drug, experienced flight to a witches' Sabbath and met Satan himself (Harner 1973). In contemporary Euro-American society, patients in psychoanalysis soon find that the content, as well as the interpretation, of their dreams changes. Dreams of Freudian analysands are different from those of patients in Jungian analysis. Had I not been so deeply immersed in Malay thought when I suffered from dengue hermorrhagic fever and its aftermath, I would not have experienced my pain and its cessation as the loss and return of *semangat*. Were I a believing Christian, I might have interpreted my vision of the birds as a visitation from the Holy Spirit of the Annunciation that appeared to the Virgin Mary in the form of a dove. The content of the vision itself might have been different: perhaps, instead of a series of birds, I might have seen Christ and the Virgin clasping me to their bosoms as they restored my health.

The dialectic between symbolic reality and bodily experience supports rather than questions paradigms. Humoral reasoning reigned supreme in Europe for at least two thousand years, and no amount of contrary evidence could have been sufficient to cast doubt on its truth since any evidence was evaluated in the light of this powerful paradigm. It took a change of paradigm to eliminate humoral pathology from the canons of Western medical thought.

We all experience empirical reality, the reality of our own senses. Sensory input is vital to the maintenance of a symbolic system; moorings to an experiential world keep the symbolic structure from detaching itself from human existence. But the "proof" that they carry is only meaningful because of the symbolic structure which allows us to interpret experience in a way that helps us believe that the cosmos itself is meaningful, that things connect, that life has an aim, and that human beings, at least to some extent, can acquire knowledge to deal with the workings of an orderly universe.

REFERENCES

Awang Sudjai Hairul and Yusoff Khan (1977) *Kamus Lenkap*. Petaling Jaya: Pustaka Jaman.
Burton, Benjamin T. (1965) *The Heinz Handbook of Nutrition*. New York: Macmillan.

Firth, Raymond (1974) Faith and Skepticism in Kelantan Village Magic. In William R. Roff, ed., *Kelantan: Religion, Society and Politics in a Malay State*. Kuala Lumpur: Oxford University Press.

Gordon, Bernard L. (1977) Fish and Food Poisoning. *Sea Frontiers* 23 (4): 218–27.

Halstead, Bruce W. (1959) *Dangerous Marine Animals*. Cambridge, MA: Cornell Maritime Press.

(1967) *Poisonous and Venomous Marine Animals of the World*. Washington, DC: US Government Printing Office.

Harner, Michael, ed. (1973) *Hallucinogens and Shamanism*. London: Oxford University Press.

Iskandar, T. (1970) *Kamus Dewan*. Kuala Lumpur: Dewan Pustaka dan Bahasa.

Kapferer, Bruce (1983) *A Celebration of Demons: Exorcism and the Aesthetics of Healing in Sri Lanka*. Bloomington, IN: Indiana University Press.

Kessler, Clive S. (1977) Conflict and Sovereignty in Kelantanese Malay Spirit Seances. In Vincent Crapanzano and Vivian Garrison, eds., *Case Studies in Spirit Possession*. New York: John Wiley and Sons.

Laderman, Carol (1983) *Wives and Midwives: Childbirth and Nutrition in Rural Malaysia*. Berkeley, Los Angeles and London: University of California Press.

(1991) *Taming the Wind of Desire: Psychology, Medicine, and Aesthetics in Malay Shamanistic Performance*. Berkeley, Los Angeles, and London: University of California Press.

Mills, Jean (1958) Modifications in Food Selections Observed by Malay Women During Pregnancy and After Confinement. *Medical Journal of Malaya* 13 (2): 139–44.

Mohd. Taib Osman (1976) The Bomoh and the Practice of Malay Medicine. *The South East Asian Review* 1 (1): 16–26.

Provencher, Ronald (1971) *Two Malay Worlds*. Research Monograph No. 4, Center for South and Southeast Asia Studies, University of California, Berkeley.

Ramanamurthy, P. S. V. (1969) Physiological Effects of "Hot" and "Cold" Foods in Human Subjects. *Journal of Nutrition and Diet* 7: 187–91.

Wilkinson, R. J. (1959) *A Malay–English Dictionary*. London: Macmillan.

Wilson, Christine S. (1970) Food Beliefs and Practices of Malay Fishermen: An Ethnographic Study of Diet on the East Coast of Malaya. Ph.D. Dissertation, Dept. of Nutrition, University of California, Berkeley.

(1971) Food Beliefs Affect Nutritional Status of Malay Fisherfolk. *Journal of Nutrition Education* 2 (3): 96–8.

Zainal-Abidin bin Ahmed (1947) The Various Significations of the Malay word *Sejok. Journal of the Royal Asiatic Society, Malay Branch* 20 (2): 41–4.

Part IV

Pain and meaning

9 Chronic pain and the tension between the body as subject and object

Jean Jackson

This chapter uses an "embodiment" approach to examine some of the mysteries of chronic pain. An embodiment perspective "requires that the body as a methodological figure must itself be nondualistic, i.e., not distinct from or in interaction with an opposed principle of mind" (Csordas 1990: 8). I approach pain as something quintessentially lived and experienced in the body. As much as possible I assume the experiencer's perspective of being-in-the-world (Merleau-Ponty, 1962 (quoted in Csordas 1993: 3)), specifically in the pain-full world. I explore the ways in which the experience of chronic pain is simultaneously sensation and emotion, neither preceding the other, and critically examine our model of chronic pain as first caused (in either the body or the mind) and then experienced.

This discussion also looks at pain as preobjective, that is, not yet involved with a subject–object distinction, and explores the connections between language and pain: the role played by our analytic language, and the notion that a pain sufferer speaks the language of a world different from the everyday world. This language, and pain sufferers' opinions about successfully communicating about their lived pain are also considered, drawing on Schutz's notion of multiple realities, in order to understand why people with chronic pain report feeling profoundly understood by fellow sufferers and profoundly misunderstood by non-sufferers.

Introduction to the research

I draw on data collected during an ethnographic research project in an inpatient chronic pain treatment center in New England (here referred to as the Commonwealth Pain Center, or CPC). The CPC, a separate twenty-one bed inpatient unit in a private nonprofit rehabilitation hospital, offers a multidisciplinary one-month program geared to reducing chronic pain and teaching skills for coping with it. Treatment involves a team approach and focuses on conservative, non-invasive therapies, including physical therapy (exercise, whirlpool, ice massage, ultrasound, transcutaneous nerve stimulation); cognitive therapies (relaxation training, biofeedback); social service;

group psychotherapy; and one-on-one psychiatric therapy. Goals of this center – like those of many other centers around the country[1] – include eliminating the source of pain when feasible, teaching the patient his or her limitations, improving pain control, relieving drug dependence, and treating underlying depression and insomnia. The CPC also attempts to examine issues of secondary gain,[2] tries to improve family and community support systems, and in general works at returning patients to functional and productive lives. The majority of patients have lower back pain. Next in frequency are headaches and neck pain, followed by complaints of facial, chest, arm and abdominal pain.

The staff sees some patients as entering with an acknowledged-by-all "real" problem with overlays of depression or "pain habits." Other patients are admitted with a problem seen as originally organic (e.g. the result of a car accident), but now including significant psychogenic elements. And many patients come in with mysterious pain problems which staff must bring out into the open (asking what patients consider prying questions) in order to arrive at a more comprehensive diagnosis. No patients are admitted with "uncomplicated" chronic pain – pain due to arthritis or osteoporosis, for instance – if the patient handles it as well as could be expected. "If a man has pain from degenerative disk disease but doesn't complain about it, is working full-time, and has a good relationship with his wife, he doesn't need external contingency management" (Schaeffer 1983: 24).

All CPC patients not only have chronic pain but exhibit chronic pain syndrome to some degree, a condition in which pain has taken over a sufferer's life. Chronic pain is defined as any continuing pain that has lost its biological function (Bonica 1976: 11; Black 1979: 34). Pain usually indicates nociception (the onset of a provoking or harmful condition) – that is, it is a symptom rather than a disease, a "normal" indication of something abnormal. Any chronic pain whose underlying pathology is not totally clear (as with most patients at the CPC) confounds the common-sense notions of disease and health. Most clinicians agree that acute pain turns into chronic pain after the somewhat arbitrary cutoff point of six months,[3] but disagree about the extent to which chronic pain is due to psychogenic, rather than physical causes, and the consequent implications for treatment.

Research was carried out at the CPC from February 1986 to February 1987, which included eight months spent on the unit observing and interacting with a total of 173 resident patients. In all, I conducted 196 interviews with 136 patients (60 of whom were interviewed twice) and interviewed 20 staff members as well. My main concern was to investigate "cognitive restructuring" in patients. The social context of treatment at the CPC, involving intensive interaction, results in patients creating new intersubjective meanings of pain during their stay. I wanted to know if these changes in the way patients thought about pain correlated with self-reports about improvement. In addition, a majority of patients resented at least some of the policies and ideology of the program, and I was interested in how patients dealt with themselves, their pain, and staff in a setting of such confrontation.

Confusion is another characteristic of life at the center. This, as I have argued elsewhere (J. Jackson 1992), is the result of (1) the nature of pain itself (its invisibility, its subjectivity, its challenge to Western mind–body dualism), (2) contradictory ideas about responsibility for one's condition, (3) stigma and demoralization, and (4) the CPC's unusual program. Pain can be communicated only by pain behavior; we understand pain behavior as either an attempt to communicate the experience of pain itself, or an attempt to express affects – suffering, demoralization, and other feelings and ideas associated with the pain experience. But I will argue here that separating "the pain experience" from other experiences accompanying pain, somehow viewing it apart from "real" pain itself, is an impossible task. From a phenomenological perspective, the distinction between pain behavior, the experience of pain, and the emotional states accompanying pain, is highly ambiguous. Clearly, intractable chronic pain is a condition involving a great deal of suffering, an existential affliction involving bodily, mental, emotional, and spiritual distress.

Subject–object

Phenomenology wants us to begin analysis with the "lived world of perceptual phenomena" (Csordas 1993: 23). In this world neither our bodies nor our pain are objects. But in analyzing an interview with a chronic pain sufferer, Good notes that Brian was objectifying his body. Rather than "simply live through his body in the 'world of everyday life,'" (Good 1992: 38), Brian's body was being taken over by pain: "the pain has agency. It is a demon, a monster ... Pain is an 'it'." (Good 1992: 39).

I encountered many similar examples. In part, as we shall see, one must objectify one's body or one's pain when using everyday language, but the strictures of linguistic interaction do not entirely explain this process. Pain confounds a simple subject–object dichotomy because objectification and subjectification stand in a dialectical relationship to each other. Brian's account illustrates the way he objectifies both his pain and his body: "when I think ... I'm outside myself ... as if my mind were separated from my self, I guess." (Good 1992: 39).

An example of subjectification, on the other hand, would be the patient who asks if the pain is "all in my head." "All in my head" pain, since it is a figment of the imagination, cannot be an object "out there." (Note that "out there" can refer to inside the body; what is important is that pain is seen as an object.) Patients sometimes say – although usually when joking – they would welcome this kind of subjective pain because, even though "all in my head" pain is stigmatized pain, subjectifying the pain by accepting that "it doesn't exist" should make it less horrible. Brian, for example, wants to "explain it away ... say that it's all just imaginary, it's a figment, it really doesn't exist" (Good 1992: 40). One CPC patient's comment suggests the allure of this possibility:

hearing that if my mood were better, maybe the pain would go away. I tried willing it away ... "Well," I figured, "if I did it to myself, I can get rid of it." And I kept telling myself, "I don't want you, get out of here." Didn't work.

In fact, for most sufferers, subjectification does not produce peace of mind. First, many (like the patient quoted above) find they cannot "wish it away." Second, if something so horrible is imaginary, then one must be seriously delusional.

Understanding the subjectivity of pain is not simply a matter of establishing the equations "object = 'out-there-and-real,'" and "subject = 'not real,'" because experientially the pain remains very "real" for the sufferers. Rather, the subject–object dichotomy is essentially about the presumed cause of a pain, and while this is surely an important factor in how one experiences pain, it is not the same thing as the experience of pain itself. When Brian tries to say "the pain doesn't exist" he is trying, like many pain sufferers, to use language to escape from his current experience of pain via changing its meaning by attributing a different origin to it. Unfortunately, sufferers usually report that this linguistic (de)representation "didn't work" – their lived reality continues to be pain-full.

A very striking discovery in my interviews was that pain patients seek relief both by moving toward increased objectification of pain and moving toward increased subjectification of it. Issues of control lie behind this straddling of the subject–object boundary. Thus those who move toward greater subjectification do not attempt to decrease the power of their pain by saying it does not exist, but rather attempt to merge the pain more with their selves. The crucial element seems to be how pain sufferers connect pain to their bodies and to their identity. If one claims it, accepts it, does not fight it (even, at times, identifies with it), then, paradoxically, à la the lesson of a Zen master, one better controls it:

PATIENT: It's how you perceive the whole thing, how you look at it. If you look at the pain as being a horrible, terrible thing, then it is going to do horrible, terrible things to you. If you look at it as being a netural thing – it is just *there* – it loses a lot of its power.

JJ: Sounds like you are more accepting of the pain.

PATIENT: Yes. I have accepted. I thought I had pretty much accepted it before – oh, a couple of years ago – except for the fact that I was always going to be in pain. But there was always in the back of my mind, saying, "But maybe one of these days I'm going to go to a doctor and they're going to do some X-rays, they're going to do a test and they're going to say, "Ah-hah! Here it is, here's a bone chip." It's kind of a fairy tale of mine that some day, wave a magic wand and make it all better. Uhm, now I don't really care, you know, I don't think that day will come. And the acceptance that that day won't come means, so it's going to be with you, so why not deal with it?

In contrast, patients who opt for increasing the objectification of pain speak in more conventional terms of "getting a grip on it":

and if I ever were to give in to the headaches, completely give in, I'd be an invalid. Will I win, or my head?

You have to learn to listen to your body, but then again you can't let your body dictate everything to you, there's a fine line.

And I'm not going to give it an inch. I'm going to keep pushing it and who knows? Maybe in a year, maybe in ten years, maybe never, but maybe someday I'll have pushed it back to the point where I don't feel it, where it is all not felt.

Some of the explicit therapy at the CPC involves encouraging patients to objectify their pain more, for example, by pain imaging. After these sessions, patients readily speak of their pain as sea serpents, crabs, medieval weapons, and the life. Presumably this sort of objectification allows a patient better to "understand" the pain, i.e., to gain therapeutic insight into its cause. For example, one patient reported becoming aware that the creature in her mind's eye resembled her father. Some patients report that objectification brings greater acceptance of their pain – such as a patient who reported that although his pain was an ugly black monster on the sea floor, he found he really didn't want to kill it. At times an increase in subjectification seems to follow the initial increase in objectification: after imaging the pain as something entirely separate from the body and self, one makes some kind of identification with the pain, e.g., that the pain is one's father. Although still aversive, the pain is more accepted since it is a part of oneself.

Another patient reported that whereas earlier she had seen her pain and her back as "foreign" objects, she learned through imaging that her pain was still "me" after all:

JJ: Is the snake you or not you?
PATIENT: I believe the snake is me ... "My God, I can talk to it." I was afraid of it. And, plus, I have these two knot-rods in my back, and to me it was something foreign in there, so it was very easy for me to put together something in my back, because I do have something in my back that is foreign. But the pain is me.

These remarks – and there are many more like them – reveal that we can draw no simple conclusions about pain as either subject or object.[4] Pain is not unique in this respect: people speak of many kinds of feelings as being both "me" and "not-me." In poetry about being in love, it is the poet's heart that is pierced, the poet's soul that has found bliss, yet we also find many descriptions of feeling possessed or of merging with the beloved. I will argue below that being profoundly in love, like being in severe pain, requires leaving the everyday life-world; in the alternate worlds of over-whelming love or severe pain the self is transformed such that "me" and "not-me" converge in some ways. The difference is that a journey to the province of love is willingly undertaken, at least some of the time, whereas pain animates the sufferer to return to the everyday painless world.

Brian is frustrated in his attempt to objectify both his body and his pain both because pain resists the objectification provided by standard medical testing such as X-rays (Good 1992: 39), and threatens the objective struc-ture of the everyday world in which he lives. Good (1992: 39) notes that:

We act in the world *through* our bodies; our bodies are the subject of our actions, that through which we experience, comprehend and act upon the world. In contrast, Brian described his body as having become an object, distinct from the experiencing and acting self ... At the same time, pain is a part of the subject, a dimension of the body, a part of the self.

Subjectification can increase as pain progressively destroys the world. Experiencing severe pain, one becomes one with the pain, one becomes pain. That this is an unwelcome merging (unlike, for instance, a fervently sought merging with the Holy Spirit) does not negate the fact that the subject, the me, moves towards becoming one with the pain. One can be terrified of this process, like the CPC patient who spoke of becoming "psychotic with pain." Or one can welcome an increase in subjectification of pain: 'It's like I know it's a part of me now and it really doesn't have to be as bad as it is if I work with it instead of against it, if I really integrate it into my day.'

However, the usual response to pain involves attempts to objectify it and separate it from the self. Experiencing a mild and pleasant trance state, we would likely make no attempt to find its cause or control it, but we do try to objectify pain in order to understand and control the difference between painful and painless states. We are also likely to think of pain as an object because it can be so powerful and can produce such grave and aversive effects. Something that can virtually obliterate consciousness of anything but itself, that can "destroy the world" (Scarry, 1985) is clearly an "it," existing apart from the self in some fashion.

As members of present-day secular Western culture,[5] most of us tend to think of pain as something physical: since we experience pain in the body, in so far as we see our bodies as physical objects, so will we also see pain as physical. We understand the origin of pain to be physical as well: touching fire with a finger initiates chemical and electrical processes, and pain results. Pain itself is sometimes conceptualized in terms of the presence of a physical object inside oneself. I collected many statements from both patients and staff characterizing pain in this way: pain not only exists and is real, one can see it in an X-ray or CT scan.[6] Again, the presumed origin of a pain (a ruptured disk) is conflated with experiencing it for both professionals and sufferers, although professionals will also talk about the difference between cause and experience – when, for example, examining an X-ray suggests there should be massive pain and yet no pain is reported.

Finally, in a clinical setting one will be inclined to think of pain as physical, for pain spoken of as non-physical is suspect (see J. Jackson, 1992). Further discussion of the issue of physical causes of pain versus the physicality of pain itself follows below.

In short, chronic pain challenges the notion of the body as object and the self as subject. While the subject can be the conscious mind "having" an objective body and an objective pain, it is also true that the subject can combine with pain, becoming the "pain-full me" (perhaps contemplating a past or future "pain-free me"). As Good (1992: 45) asks, "Is the pain an essential part of the self, or is it 'merely' a part of the body? Can the pain be separated as object from the self as subject, thus differentiating the subject from the world which acts upon it, or must he 'passively endure' the pain?"

Although we cannot say that sufferers themselves consciously seek to transcend body–mind and subject–object dualisms (as we shall see, they are quite invested in maintaining mind–body dualism and, at least some of the time, subject–object dualism as well), their talk about pain allows us to examine these dualisms and better understand how the lived pain-full body confounds them.

Mind–body

For the most part chronic pain patients see their problem as one of "matter over mind" because their intractable pain makes them feel that their bodies are powerfully influencing their minds. Some go so far as to use an idiom of possession, saying that their bodies are taking over, are driving them crazy. Many of these people have found that their lives, their emotions, their spirituality, their personalities, their destinies are dominated by their pain-full bodies, and their role as sufferer has shunted aside previous roles as caregiver, provider, lover, companion, parent, friend, citizen.

Most CPC patients resist what they hear as messages about "mind over matter" from staff and any number of other sources. Such messages say that one can use the mind (for example, through biofeedback training, or various relaxation or imaging exercises) if not to gain mastery over the body to the point of eliminating the pain, then to learn to cope, to push pain aside, to go beyond it. "Get on with your life" is a frequently heard phrase at the center: will and knowledge can conquer the body; or, as it is sometimes put, will, acceptance, knowledge, and love can work with the body in a partnership to diminish the amount of suffering caused by the pain. No matter how indirectly the notion is phrased, the ultimate message is one of control, of mind over matter. A favorite phrase at the CPC is, "before, my pain controlled me, now I control my pain." The following remark is illustrative:

I was hoping to leave with a decrease in pain, but I haven't. But because of what I have learned about myself – I don't know if it makes sense – ... there is less felt pain. It's how you perceive the pain, how I perceived the pain. When I came in, the pain controlled what I did.[7]

The stark contrast of "matter over mind" and "mind over matter" illustrates our legacy of Cartesian dualism. Neither adequately describes the lived experience of patients in severe constant pain nor their accounts of how they alleviated it. A careful look at how patients talk about pain shows it is very difficult to classify their attitudes in terms of the "mind over matter" or the "matter over mind" discourses. I submit (and argue more fully below) that pain confounds this dualism because it is not, as we tend to see it, a stimulus leading to sensation accompanied by overlaps of emotion and aversion. These three components cannot co-occur simultaneously in a polar mind–body model that postulates a linear, chronological process involving origin in either body (e.g., acute pain resulting from a burn) or mind (e.g., psychological predisposition) and subsequent experience.

It is true that the vast majority of CPC patients see their pain as primarily "real pain" – physical, body pain (see J. Jackson, 1992). Even migraine patients who attribute the onset of their pain, in any given situation, to emotional stress see the resulting pain as physical, an effect of vasocongestion. Here the subject feeling the pain talks about pain as an object. The body, another object, has pain.

Yet this is not the entire story, for pain has not yet become object in much patient talk, and their discussions allow us to examine how mind–body and subject–object dualisms break down in accounts of lived painful body experience. As noted above, this is not because the sufferer denies mind–body dualism: indeed, much of the time he or she is fighting to have the bodily components of the pain recognized and acknowledged to a greater extent than they are.

Patients speak of (and, therefore, we can posit, actually experience) their pain at times in terms of an identification of self apart from pain, with the pain as alien, an intruder, an invader; at other times they use terms revealing an identification of self with pain, or pain experienced as coterminous with the body – the pain-full body, which is, at least intermittently, coterminous with the subject. Knowing where a sufferer positions himself or herself on this continuum helps us understand how an individual experiences the body and how the pain-full body has determined the self these individuals have acquired. Although unwanted, rejected, and alienated, this new self is, like the new body that accompanies the self, still one's self. In extreme cases, at times phenomenologically one *is* pain: one's selfhood and one's body combine with pain. And all the time pain is a major component of the new self, the new identity.

According to phenomenology, in the everyday world we do not normally experience our bodies, nor our pain, as objects (see Leder, 1990: 69–102). It is when we try to pay attention to pain or to talk about it, to "make sense" of it, that we objectify it. Experiencing severe pain, we simply are "in pain," we are "pain-full":

if I say that my foot hurts, I do not simply mean that it is a cause of pain in the same way as the nail which is cutting into it, differing only in being nearer to me ... I mean that the pain reveals itself as localized, that is constitutive of a "pain-infested space." "My foot hurts" means not "I think that my foot is the cause of this pain," but "the pain comes from my foot" or again "my foot has a pain." (Merleau-Ponty 1962: 93)

Pain is, in Schutz's terms, *behavior* or *conduct*, because it is a subjectively meaningful experience.[8] Pain collapses Schutz's distinction between *mere doing* and *mere thinking* (1971: 211) because pain is by definition simultaneously bodily experience and mental–emotional experience. By consisting of both sensation and emotion, by being simultaneously thinking and doing, pain confounds mind–body dualism. While we do not normally think of pain as emotion, I can find no definition of pain in either clinical or philosophical literature that does not include the notion of aversiveness, and think it impossible to construct a definition without this element.[9] Therefore, we must conclude that pain *is* emotion, not just something that has an overlay of emotion (compare Jenkins and Valiente, Chapter 7 of this volume). Where pain is deliberately sought – as in "no pain, no gain" aerobic exercise, or as a route to sexual or religious ecstasy – it is welcomed as a means to an end in spite of its aversive qualities; in these cases, pain itself can be spoken of as desirable because of a learned association with something desired. The point is simply that pain accompanied by no emotion would be psychologically pathological; in general, whether we are speaking clinically, in terms of biological function, evolution, or experience, pain is aversive.

Schutz asserts that pain is "simple" experience, like an eye blink, because he is thinking of pain as a sensation. It is true that the proximate causes of pain can be very similar to the causes of eye blinks – electrical impulses in the nervous system. But, again, the cause of pain is not the same as pain itself – the experience of pain – just as what makes pain disappear is not the same thing as the absence of pain. This seems quite obvious, and yet when we distinguish between "physical pain" and "emotional pain," as we often do, we mean by "emotional pain" either pain with no embodiment (and hence different from "physical pain"), or we are indeed differentiating the two in terms of cause. "Pain" can be used metaphorically, and I would argue that speaking of a disembodied emotional pain is an example of just such usage. But any pain experienced in the body is experienced in the body, period, and will be experienced as physical pain, regardless of the cause.

Knowing the cause of a pain is obviously often important. For one thing, such knowledge carries implications for choosing an appropriate treatment program. Also, since Western culture places such emphasis on this distinction, someone who believes his or her pain is due to a physical cause (e.g., angina produced by hardened arteries) will probably experience pain differently from a pain he or she believes to be emotional in origin (e.g., physically experienced heartache caused by unrequited love[10]). Pain sufferers pay attention to this dichotomy because of the stigma that accompanies a purely emotional pain that does not go away. A lover's heartache will become negatively evaluated (romantic literature notwithstanding) if time fails to diminish its force: the sufferer should "get over it" after a time. A clinician would diagnose such a pain in a way the sufferer would probably see as stigmatizing. Thus, the sufferer's belief about the cause of the pain will usually have an effect on the experience. But the cause of pain must be kept conceptually distinct from the embodied experience of it.

The cultural meaning of pain for chronic pain patients

Since all pain is subjectively meaningful experience, all pain has cultural meaning. One anecdote used at the CPC to illustrate this argues that someone who sprained his ankle winning a soccer game will experience the pain quite differently from someone whose sprained ankle results in the game being lost. Merleau-Ponty sees perception as preobjective, but would not contend that this means it is precultural (Csordas 1990: 9–10). Phenomenologists like Merleau-Ponty dispute the objectification of the body that occurs in a natural-science approach to human behavior; in general terms, phenomenologists speak of the body as the existential ground of culture (Csordas 1990); and culture is seen as a projection of the body into the world. In these terms, the chronic pain sufferer must create a new projection from his or her painful state, and this new state literally incorporates culture – since all pain has meaning given by the cultural world one lives in. There simply is no pain that is strictly biological;[11] pain always has meaning, always is "socially informed" (see Csordas 1993: 3; cf. Bourdieu 1977). Thus rather than speak of pain as physical sensation with overlays of meaning, we need to speak of pain as permeated with meaning – permeated with culture.

People in chronic pain struggle with its meaning partly because they have been socialized to see pain as a sensation, as basically a physical feeling. Even Schutz presents pain as something happening to someone, something undergone or suffered passively, and pain patients find themselves needing to speak of their pain in similar terms, because the mind–body distinction requires that pain be spoken as only sensation – albeit with overlays of emotion.

This interpretation produces additional suffering for many individuals. Hilbert (1984) maintains that unending pain simply cannot be comprehended as a notion and that chronic pain sufferers have "fallen out of culture." People who do not suffer chronic pain can perhaps grasp its meaning intellectually, but they will lack a deeper, empathic comprehension. Chronic pain sufferers can find themselves in a state of being that has almost no meaning for non-sufferers: if pain is primarily sensation, how can one have a never-ending sensation that takes over one's life? Furthermore, most non-sufferers seem strongly to resist the notion that one can be in severe, unending pain; it is too threatening to contemplate.

Evidence of this difficulty comes from chronic pain sufferers who report that they find communicating about their pain – explaining its cause, describing their experiences – a more serious problem that the pain itself. One reason is the expectation that automatic behavior, such as cries or grimaces, will accompany the sensation of pain – and the further expectation that such behavior will stop, sooner or later. Hence, chronic pain sufferers can be baffled about how to communicate their pain, and their own cultural expectations regarding pain do not help: some moan and groan constantly and are negatively reinforced because such behavior is acceptable only temporarily; those who exhibit no overt behavior at all have a problem because many people find it hard to believe someone is experiencing severe pain if they do not cry out, grimace, or complain. Thus those who opt for "suffering with dignity" must still find a way to communicate about the pain periodically if they want to be believed; otherwise their pain is forgotten or diminished in importance by those around them.

Some patients whose cases are not clearcut assert that they would far rather have a known problem:

Then [the doctor] said, "Rosalie, you absolutely have a problem." And I says, 'Oh, Thank God it is not all in my head." And he looked at me like I had four heads.

PATIENT: Sometimes I wish there *was* something [wrong].
JJ: Why would you want this?
PATIENT: So we could work on something, a particular problem. If it is my hormones, correct it, a chemical imbalance ... find something specific. My pituitary. A vein, or vessels are too small? Widen it.

It gets to the point when you go for something [e.g., a test] you actually hope they find something.

Well, in this day and age everyone is always afraid that if they find something, it could be cancer or it could be heart ... and no one wants that but you just hope they find something. You even get to the point you don't care what it is, no matter what it is. The fact that they found something.

I'd rather have cancer. I really would.

Clearly, in patients' lived reality pain is far more than sensation. But the sufferers themselves are reluctant to see pain as something complex because this diminishes its "real pain" quality. Even pointing out that one must be conscious to experience pain smacks of "psychology" to some patients, who are all too aware that their problem could be explained in terms of mental illness or some kind of character flaw. My interviews illustrate quite sharply that patients protest loud and clear at any hint that a given pain is "emotional" and therefore not ultimately produced by a physical cause.

During an interview, most patients will move from describing their pain in terms of sensation produced by physical causes to more global concerns, such as telling how they feel about their current, transformed, pain-full selves. But many find talking about such comprehensive issues difficult. Their tendency to stress the physical components of their pain derives in part from their struggle for legitimacy, and in part from difficulties finding an adequate language with which to talk about these issues.

Pain and language

Chronic pain sufferers often say they have trouble communicating about their pain, especially severe pain. Some part of this difficulty seems to reflect a contrast we make in which language belongs to the mind, not the body. True, much of language can be traced to the fact that we speak and think from an embodied perspective (see Johnson 1987), but some aspects of this embodiment are favored over others: for example, visual perception is encoded, at least in English, far more than aural or olfactory perception (see Rorty 1979; M. Jackson 1989). It is significant that we use the expression "body-language" to describe someone conveying meaning with his or her body. Body language is often thought of as "primitive" or "pre-linguistic" language.

Pain is also seen as pre-linguistic, an unpleasant sensation virtually without additional meaning. Our model of pain as sensation presents it as something that just *is*. Of course, with information about its mechanical, chemical, and electrical causes one can begin to treat pain linguistically, give it a name, and characterize it briefly. With further information, for example, about reactions of the rest of the body, mind and emotions, a given type of pain can be handled in a more linguistically expanded fashion. But without such information, actual pain will resist verbal description because we see it to be of the body and therefore pre-linguistic – like such other sensations as odors, music, or inner states such as hunger or sexual arousal. What distinguishes severe chronic pain is its extreme aversive quality and its persistence despite efforts to end it.

Another reason pain resists language is its invisibility. Pain cannot be measured, cannot be ascertained apart from the sufferer's affirmation of its presence. Scarry (1985: 4) writes, "pain comes unsharably into our midst as at once that which cannot be denied and that which cannot be confirmed."

Of course when we try to talk about it, pain will automatically acquire some subject–object distinction and so lose some of the quality of pre-objective, "pure" pain. Neither Merleau-Ponty himself (with his insistence that "my foot has a pain" (1962: 93)) nor Schutz can avoid the fact that since we must use language in our analyses, we necessarily create the subject–object distinction. Merleau-Ponty (ibid.: 140) states "I am not in space and time, nor do I conceive space and time; I belong to them, my body combines with them and includes them." He would similarly have to say something like "I combine with pain, I include pain." The awkwardness of expression reveals the difficulty of using everyday language to describe the experience of pain even when we want, as does Merleau-Ponty, to constitute our analyses around the preobjective act of perceiving pain.

Patients, too, necessarily objectify pain when they talk about it and even those who, like Brian, can speak eloquently on the subject frequently comment about how difficult it is to talk about some features of pain. Pain sufferers often report that only other pain sufferers understand their pre-objective, pre-abstract experiences of pain, but not through the normal medium of communication – everyday-world language. Patients claim that other forms of communication, intuitive and involving a kind of *communitas*, facilitate mutual understanding:

[I wanted to be in a place] . . . where I would be surrounded by a group of people who I would not have to explain my pain to. And that is exactly what I found.

A pain-full body occupies a world different from the everyday world. Just like the ineffable worlds of dreams, day-dreams, or deep religious or musical experience, the pain-full world has its own system of meaning and the inhabitants have their own forms for communicating that meaning. We can say this world has its own language, its own cognitive-affective style. Kirmayer (1992: 1) correctly opposes the order of the text and the order of the body; if we can speak of the body having its own language, we can understand that the pain-full body, in a pain-full world, will speak the language of that world. This helps explain why pain "resists language" – it resists everyday-world language. Entering the pain-full world requires making "a radical modification in the tension of our consciousness, founded in a different *attention à la vie*" (Schutz 1971: 232). Just as we can remember *having* emotions or other kinds of feelings, but cannot re-experience the emotions or feelings themselves, so also can we only remember having pain, but cannot remember the pain itself because that would require going back through the looking-glass into the pain-full world and speaking the language of that world.

Although everyday-world language has trouble describing the lived experience encountered in any of these worlds, if it is allowed to metaphorize pain, then it becomes adequate to the task – indeed, often eloquent. The metaphor of "seeing stars" when subjected to a sudden, intense pain is an example. The metaphor does not describe the actual experience, for: "What is meant by 'seeing stars' is that the contents of consciousness are, during those moments, obliterated, that the name of one's child, the memory of a friend's face, are all absent" (Scarry 1985: 30).

In such pain contents of consciousness have been obliterated, but one does not lose consciousness. Communication, embodied communication, is clearly taking place but a communication so different from everyday communication that the two are totally incommensurable. The experience of "seeing stars," stripped of this metaphor, illustrates what I mean by "the language of pain." We might call it "anti-language," in that it is antithetical to ordinary natural language; but it is a code, and it communicates something.

This is not to say that pain cannot obliterate everyday language; it can and does. It is in this sense that we can speak of pain as "pre-linguistic" with "linguistic" referring to the possibility of a fully developed everyday-world language. This is not the same as saying pain obliterates meaning. Pain always has meaning, from the moment of its appearance, just as the body always has meaning (and its own language). But since these meanings differ from those assigned by everyday-language, there is a sense in which we can indeed speak of the experience of severe chronic pain as "meaningless," as Hilbert (1984) does.

Moving from what I am calling the pain-full world to the everyday world entails a shock (see Schutz 1971: 233). Yet to some degree, living in chronic pain requires a constant traveling back and forth between these worlds, for example, when a sufferer has to concentrate on another person or a demanding task and is momentarily distracted from experiencing the pain in its full force. However, when pain is severe enough, as Scarry so aptly describes, one does not easily cross over to the everyday world. I emphasize back-and-forth movement because, as noted above, experiencing severe pain is not the same thing as losing consciousness – despite most pain sufferers' desperate longing, at times, for unconsciousness. What one loses is one's normal occupancy of everyday reality. Thus, although at any one moment one can be in the everyday world or the pain-full world, if we consider many of these moments strung together over time we can see the potentially grave consequences for one's experience of *durée* (Schutz 1971: 215–16). Good speaks of the dissolution of "the building blocks of the perceived world – time, space" (1992: 41). We can add to this the dissolution, or at least undermining, of the self, and of what we can normally expect from others, such as their validation of our experience.

As mentioned above, the pain-full world is unique among non-everyday worlds in that pain is almost always aversive, although nightmares or unwelcome compulsive fantasies in the dream and fantasy worlds Schutz discusses would also have this quality.[12]

In my conversations with chronic pain sufferers I have felt that they often do in fact complain about the way pain "resists objectification." But since they must talk in everyday-world language, they speak, at best, with difficulty – as the remark on p. 208 illustrates, in which the speaker wonders if he is making sense.

Another source of difficulty is that patients have been led to believe that the physical cause constitutes the intrinsic meaning of their pain. This makes it risky for them to talk about the connections between pain and consciousness. Since any mention of emotional or cognitive solutions threatens the legitimacy of a given pain by diminishing its physical quality, such suggestions are often resisted. And pain sufferers are not mistaken in this assessment, as serious consequences *can* result from establishing that a pain is somehow not physical (see, e.g., Corbett 1986). Notwithstanding these potential consequences, many patients occasionally do try, as in the remark on p. 208, to speak of pain in other terms. But for the most part their talk is premised on their "real," their physical pain.

Scarry (1985: 11) maintains that physical pain cannot easily be talked about, whereas emotional pain can, and that physical pain is pre-linguistic, but emotional pain has meaning. It is not clear whether she is drawing a distinction between pain felt in the body in terms of physical versus emotional cause, or a distinction between embodied pain, regardless of cause, and pain felt only in the mind.[13] As she bases her discussion on physical torture, it would seem that by "physical pain" she means pain produced by physical causes. If this is so, her dichotomy is not useful for any analysis dealing with the lived reality of pain: phenomenologically speaking, emotionally caused pain, if it is experienced physically, is physical pain.

Most patients comment that at the CPC, "for the first time" their pain is understood, adding that their fellow patients understand far better than staff:

These people [staff] who are trying to help, they read in books, but they don't know, because they don't have pain ... what you're going through.

The staff understands only intellectually, the community at gut level.

Patients are much better. First-hand experience, that makes a big difference. Yeah, [staff] works with it, I'm sure they can detect it by ... how other patients have reacted. I still think patients are a better judge.

The staff ... not having the pain ... they're not in the same place that you are ... not feeling the same way that you do.

In a similar vein, Brian commented on non-sufferers: "If they don't suffer with it, they don't understand it. And they're really skeptical. They don't believe in you ... Maybe this is just something you're making up" (Good 1992: 40).

Many CPC patients report something akin to joy at being in a community of fellow-sufferers like the CPC, where all members meet the preconditions for understanding *bodily-lived* meaning. There is a sense of shared experience: "And so, I began to say, 'these people feel exactly like I,' [they had also felt that] they were the only ones in really bad pain." Also shared is a sense of shared identity: "You're a member of a very exclusive club." "[A]nd a lot of closeness that happens when two birds of a feather flock together." Patients report feeling comfortable because of what they share:

I feel comfortable here because everyone's in pain.

I think it's comforting to be with people for a change who also have pain.

So one feels very comfortable and very safe. Both inhibitions and pretensions fall by the wayside. One becomes very direct, a lot of bullshit goes by the board because there's a common denominator in pain, a sense of loss.

This shared identity and feeling comfortable makes it easier to share one's suffering:

Yesterday we had focus [group therapy]. I liked it very much because there were other people who'd gone through the same thing. I broke down and cried, but it was a release.

PATIENT: It's [the community] not phony, it comes from inside.

JJ: Why?

PATIENT: Because they feel the same way I feel and it's easier to share it with somebody like that who knows what you're going through, and what they're going through. And it's a lot easier, it makes you more comfortable.

A stronger bond is forged than with others ... you have a common denominator. Another guy in focus group for the first time, it was also my first time, he said, "You're talking straight out of my heart."

Patients describe what it is like to talk to someone who seems to understand:

When you come in here, they start talking to you. They may have different pain, but they all understand, which your family doesn't.

In the outside world you don't want to talk about yourself, you don't want to hear about others. Here are people who are willing to listen.

In a regular hospital, I don't talk to anyone, I'm shy. In here, we're basically in pain, they understand.

Because they're all going through the same thing, they've been there, they can understand what my best friend can't understand.

Yet patients are quite capable of disagreeing with one another about how to interpret pain. Although such disagreements usually occur behind one another's backs, occasionally patients confront one another. I would argue that there is no contradiction here: when patients disagree with one another, they are disagreeing about the cause of pain, and therefore what treatment program is to be followed. When patients effuse about how well they understand one another's pain, they are talking about the phenomenological experience of pain.[14]

Paradoxically, pain – that quintessentially private sensation, experience, emotion – depends on social action to make it "real." In general, when a pain sufferer communicates about pain he or she intends to do so.[15] Those sufferers who do not intend a communication can quite successfully hide the fact that they are experiencing pain.[16] Pain is communicable only through pain behavior, verbal or nonverbal, and questions concerning to what degree communicating one's pain is a conscious choice – or no choice at all (i.e., a purely physiological response) – form the basis for a good deal of research and a good deal of discussion at the CPC. What is important here is that in general the body does not communicate chronic pain to others without conscious intention – with perhaps some exceptions in cases of extreme pain and borderline consciousness. Cries and screams under full anesthesia are not examples of pain being communicated. Scarry's examples of unbearable pain are horrible because extreme pain is being experienced by a conscious sentience, a sentience that would welcome unconsciousness but cannot obtain such relief.

While the everyday world is the basis of our experience of reality, and all other worlds are modifications on this basic pattern (Schutz 1971: 233), the world of severe constant pain seriously threatens that everyday world. Pain "shatters" language (Scarry 1985: 5). Moving between these worlds involves a greater shift than, for instance, moving to the world of temporary daydreaming – or the world of temporary acute pain. This is why so many pain patients feel so profoundly misunderstood by their friends and relatives: "He thinks my back pain is like his morning backache, and I can't convince him it's totally different." This patient has difficulty in part because "any language ... pertains as communication ... to the intersubjective world of working" (i.e., the world of everyday life) and hence that language "obstinately resists serving as a vehicle for meanings which transcend its own presuppositions" (Schutz 1971: 233). Brian's pain, which "shapes his world to itself, resisting objectification and threatening the objective structure of the everyday world in which Brian participates" (Good 1992: 136) also illustrates this. The pain-full world is one others "could not possibly understand," one that cannot be described except with its own language, its own morphology and syntax, and its own speech community. Attempts to translate the language of the pain-full world are understandably never very successful – traduttore, traditore.[17]

Significantly, not only do patients say that it is hard to talk about pain, they also say that once they have been at the CPC for a bit, "we don't talk about our pain." Talking is superfluous, for somehow the communication has already taken place:

We have something in common, it's sort of a bond between people. No one says anything to complain, not any more.

[Speaking of two patients who are very badly off]. But they never complain about the pain. You *know* because you're feeling it yourself.

Everyone has a disability and there's very little crying on shoulders.

You know, misery loves company sort of thing. And the strange thing is, [the more] time you spend with your fellow sufferers, you find the less you talk about the actual pain with them ... you know that they know what it's all about.

Nor, according to some, is nonverbal pain behavior necessary:

Also, there's no wincing, no facial expressions – the pain behavior. You know what I mean, people holding their backs. There wasn't any of that. [B]ut here no one tells you not to do that, none of the patients tell you not to do it.

We do a lot of laughing, but everyone's hurting.

Here's these people, they're all in pain and they're having terrible stories and they're walking around. And a lot of people say when they walk in, they can't believe these people are in pain.

What pain sufferers at the CPC have learned is that the normal medium for sharing – everyday-world language – can be a serious stumbling block to communication. Patients feel that language is inadequate, or the hand-maiden of the professionals. Brian points out that "they'll have their own answers and solutions that don't jive with your own" (Good 1992: 40). But with fellow patients: "Whatever you say about the pain is believable, just because of your common bond and their knowledge; whether it's in the same spot or not, they truly understand."

Now, a clinician might interpret this as evidence that patients are improving, that perhaps the decrease in pain behavior is due to there being less pain. Or perhaps patients are distracted and less depressed, or perhaps such behavior decreases simply because fellow patients do not reward it. Evidence can certainly be marshaled for the latter interpretation:

And it's funny, when you've been here a while you notice the new people will show those behaviors. But they get slowly extinguished because no one responds – and I guess maybe you feel embarrassed or something. You certainly become aware of that behavior when you're around other people in pain who aren't doing it ... it's very clear that since no one else is doing it, it's really not OK, it's not really accepted.[18]

It is clear to me that this observation is accurate (despite the comment quoted above: "no one tells you not to" engage in such behavior). But there is an additional reason for patients not communicating verbally or nonverbally about their pain after a week or so at the CPC: since they report difficulty communicating about pain[19] in ordinary language, and since the message patients mainly want to communicate ("I am in pain and I realize you are in pain") is sent and received early on in a patient's stay, they feel no need constantly to talk about their pain for the purpose of validation. They know from experience that a major feature of their pain is its preobjective and pre-linguistic quality, so attempts to communicate further information about it will be frustrating and unsuccessful. Many patients comment about how both verbal and nonverbal resources are neither adequate for what they want to communicate, nor necessary for feeling understood: "they know when you don't grimace that you aren't necessarily free of pain."

Is non-linguistic pain behavior best understood as language, or gesture? As a search for meaning, or escape from (inauthentic, imposed, inadequate) meaning? Perhaps chronic pain patients, with their mysterious maladies, so dependent on the medical establishment to bring them legitimacy – to bring them a name, a diagnosis and, hopefully, a treatment for their affliction – have special reason to be suspicious of the language available. At the CPC, discourse about pain is either a resolutely medico-scientific discourse, as heard in the CPC's public medical rounds sessions, or a psychotherapeutic discourse, as heard from staff when leading encounter groups.[20] But the approaches represented by the medical and psychotherapeutic language, although for the most part granted authority by the patients, have not worked for them. They are all too aware that they are "the failures of the medical system," as both staff and patients periodically point out.

Glossolalia, "speaking in tongues," has an authenticity, an immediacy, a presence similar to the "cries and shrieks" – or the at times very expressive utter silence – of the chronic pain sufferer. As Csordas (1990: 27) notes: "What better way to maximize the gestural element of communitas, and what better way to preclude the petrification of *parole* into *langue* than to speak in tongues, always a pure act of expression and never subject to codification?"

This points to the dilemma noted above: although a phenomenologically based analysis should begin with the "lived world of perceptual phenomena" (Csordas 1993: 23), a world in which neither our bodies nor our pain are objects, we are still dependent on patients' talk about their experience of pain, and this talk uses everyday language. How to record the language of pain itself? Even tape recordings of "pre-linguistic cries and shrieks" (Scarry 1985: 43) would involve some objectification; something like the McGill Pain Questionnaire (Melzack 1975) requires much more.

I have concluded that pain, in a sense, *is* a language, and that it competes with everyday-world language. Scarry's discussion of language destroyed by severe torture is at one extreme location in the world of pain. For chronic pain sufferers, the messages their pain sends replace everyday-world language, transform it so the messages of everyday language – whether sent or received – become distorted, bizarre, pointless, or trivial. Having made the effort to leave the pain-full world temporarily – say a sufferer has, with considerable exertion, listened to a lecture for an hour – the sufferer returns to the world he or she so unwillingly inhabits feeling exhausted, resentful, even betrayed, for what he or she has just made a great effort to hear and comprehend now, back in the pain-full world, resembles Jabberwocky. And the reverse is also true: whatever form the language of chronic pain assumes, when sufferers attempt to translate it into everyday-world language – using gestures, cries, or metaphors – they say they feel a sense of failure, a sense of speaking a nonsense language, so poorly does one map onto the other.

Although we might agree with Hilbert (1984) that the idea of continually experienced pain is culturally meaningless in the everyday world, the language spoken in the pain-full world *has* meaning there. If one is totally engrossed in a musical experience, one cannot use everyday-world language to describe what one is experiencing, because the act of experiencing music relies on another language. One can report about the experience afterward, even describe, albeit awkwardly, the form and content of the music in everyday-world language, but one can speak of the *experience* only metaphorically. Similarly, everyday-world language must objectify pain in order to "give meaning" to it and this requires distorting the experience – in a sense, betraying it.

Chronic pain sufferers live out the pre-linguistic and pre-abstract meaning of pain and in this sense they understand its meaning. But since, with few exceptions, they cannot accept that meaning they – like Brian – search for an everyday-world meaning. Because they so ardently reject the meaning provided by the unmediated experience of chronic pain, they search for a name, a prognosis, something from everyday-world language that will provide an exit. They do not long to stay in the province of pain, they do not seek to continue learning its language or its customs. Despite their familiarity with the terrain and the "pain habits" they have developed over the years, CPC patients are unwilling sojourners in that province. Yet it is understandable that despite their longing to leave, having inhabited the territory for so long, they are pleased to find fellow sojourners with whom they can commune and communicate.

Language can betray. Brian and many other sufferers have come to suspect language, and with good reason. This is why, although CPC patients search for ways to communicate the meaning of their pain in everyday-world language to make non-sufferers understand and to get help from the professionals, since they have been disappointed time and again in this quest, they understandably find comfort in not having to talk about their pain. They report feeling tremendous relief at being accepted by fellow patients without having to prove anything, verbally or nonverbally. In similar fashion, people at a concert will report feeling a spirit of musical *communitas* with fellow listeners that needs no words; a feeling which, they would say, words would only misrepresent.

This, then, is one of the many ironies about pain. Pain patients do search for the meaning of their pain, at many levels. The human response to the search for meaning, "to reverse the deobjectifying work of language by forcing *pain itself* into avenues of objectification" (Good 1992: 29) is very evident in pain patients. Scarry (1985: 6) states that:

> To witness the reversion to the pre-language of cries and groans is to witness the destruction of language; but conversely, to be present when a person moves up out of that pre-language and projects the facts of sentience into speech is almost to have been permitted to be present at the birth of language itself.

But the language so born does not thrive because the patient continues to be an unwilling sojourner in the province of pain, frustrated in his or her search for meaning, for language, for names. Despite our perception of language as communicating, clarifying, enhancing, and creating (as in the case of performative language; cf. Austin 1962) experience, for certain purposes, everyday-world language is inadequate, in part because it objecti-fies ongoing subjective experience, in the process restricting, distorting, and mystifying. This is particularly true for pain, for as Scarry points out, pain resists objectification; despite attempts to communicate, one's extremely real pain remains unreal to others.

Scarry clearly is in favor of the "birth of language," for this implies surviving pain's onslaught. Kirmayer (1992: 324) similarly asks, "How can meaning and value be sustained when consciousness is constricted, degraded and defiled by pain?" But, as Scarry herself points out, "making" a world involves destruction: wounding and creating are closely related, and language plays a part in this. Pain, torture, and war destroy; humans collectively create languages which permit one person to destroy another: "the torturer asks the questions" (Scarry 1985: 28). Similarly, language can facilitate domination, as the hegemonic discourse of the CPC staff (in which, following Gramsci, we can see patients participating) often illus-trates. Language allows distance from experience, and while we may benefit from this feature of language when we are trying to gain control over aversive experiences, the language we use does not reproduce the link between the experience and the "me" undergoing it. We intellectualize with language, we see it as distinct from the body, and we privilege it, often stripped of the emotional (Rosaldo 1984: 143), over the body. Since pain is experienced in the body, is an emotion, and contains much that is indeter-minate and inchoate, it is often poorly served by everyday-world language.

Thus, in another paradox, patients both pursue language – answers, names,[21] definitions, meanings that promise reassurance and cures – and avoid it. Although they have found that language fails to represent their being-in-the-world, that promising meanings turn out to be siren-meanings, that their quest to be understood as pain-full beings remains unfulfilled, they also want to use language to escape that experience, that world. Although they report feeling profoundly misunderstood, pigeonholed, and categorized by everyday-world language, this is the language they continue to pin their hopes on.

We can say that patients both seek meaning for their pain and resist the understanding of their pain that they already possess, acquired through profoundly living it. The problem is that this meaning is unacceptable: this meaning is an abomination, it is hell itself; no one should be in constant, severe pain. Every time they speak, or groan, or remain silent, and are disappointed (at times a disappointment approximating despair) at the results of their choice of action, they compellingly illustrate the incommen-surability between embodiment-as-lived and embodiment-as-represented. They long for an adequate everyday-world explanation, provided in a language that promises distance, control, abstraction precisely *because* representation is not coterminous with experience itself.

Conclusions

This paper has looked at pain from a perspective of embodiment, which requires that "the body as a methodological figure must itself be nondualistic, i.e., not distinct from or in interaction with an opposed principle of mind" (Csordas 1990: 8). Csordas argues that an embodiment approach denies an interaction between poles of conventional analytic dualities, especially causal interaction: it is incorrect to argue either that mind determines body or the inverse. This approach requires that we not speak of physical pain as categorically different from mental pain, unless what we mean by "mental pain" is pain not experienced in the body, what I call "metaphorical pain."[22] The clinical literature distinguishes organic pain from mental/emotional pain ("real" pain versus "psychogenic" pain), but that discourse concerns causes rather than the experience itself. Experientially, we are left with indeterminacy: the image of Cupid's arrows literally stabbing the heart (since some people experience the pains that can accompany love in the chest area) or the ambiguity of "to feel" with respect to sensation or emotion. That we are most often invested in seeing pain as physical, real, and legitimate, is of course profoundly important for our analysis because all pain has cultural meaning: pain that is not experienced in the body at all (sadness with no bodily sensations, for instance), will be experienced differently from pain from an ingrown toenail. Physical pain that one believes to be derived from a psychological source (heartache from unrequited love) will feel different from pain that one believes has a material cause (angina). But establishing the experiential boundary between emotional pain and physical pain is difficult when both are experienced in the body. This is why I disagree with Scarry's insistence on drawing such a clearcut distinction between physical and emotional pain: the lines are seldom so clear phenomenologically, especially when considering pain that lasts and lasts.

When we see pain as lived, as experienced in the body, we can see it as preobjective, that is, not yet incorporating a subject–object distinction. And we can understand that pain sufferers who inhabit the pain-full world try to extricate themselves from this world by attempting to create such a distinction. But as we have seen, while some sufferers move towards increasing objectification of the pain, others move towards increasing subjectification. The reasons for this are clear: pain remains essentially preobjective, and since creating a subject–object distinction fails to remove the pain, additional strategies are attempted. Some of these strategies work; individuals who have pain but who either distance themselves from it via meditation, or try to get closer to it to experience it as less aversive, can be seen as not entirely failing in their attempt to cope with the unwanted sensations and emotions that comprise pain.

Finally, Schutz's notion of multiple realities helps us better understand several aspects of the experience of chronic pain, in particular why chronic pain sufferers report feeling profoundly misunderstood by non-sufferers, and profoundly understood by fellow sufferers. They are exiles in the province of pain, and they find everyday-world language inadequate for communicating about their experiences there. And yet, because they are unwilling sojourners, they continue to turn periodically to that very same everyday-world language to avail themselves of its promised rationality, order, explanation, and control.

ACKNOWLEDGEMENTS

An earlier version of this paper was given in the symposium on "The Body as Existential Ground of Culture" at the 1990 American Ethnological Society Spring Meeting in Atlanta. Research at the CPC was funded by NIMH Research Grant MH41787 and sabbatical funds from MIT. My deepest thanks to the staff and patients at the CPC for supporting the study in so many ways. Thanks also to the following people who commented on previous drafts: Thomas Csordas, Byron Good, Louis Kampf, Arthur Kleinman, David Napier, and Peter Solomon.

NOTES

1 At these centers, chronic pain itself is seen as a problem rather than as an indication of a problem. These new pain centers, and the language, philosophy, international organizations, journals, and meetings associated with them can be seen as an instance of the cultural construction of a disease (see also Csordas and Clark, 1992).
2 Primary gain "is the intrapersonal, psychological mechanism for the reduction of (defense against) unacceptable affect or conflict. Secondary gain is the interpersonal or environmental advantage supplied by a symptom(s)" (Bokan et al. 1981: 331). Tertiary gain involves someone other than the patient seeking or achieving gains from the patient's illness (Bokan et al. 1981).
3 Black, however, (1979: 34) classifies chronic pain as psychogenic and uses the phrase "recurrent acute pain" for chronic organic pain.
4 Good found this to be the case as well. Although he discusses Brian's case in terms of an example of objectification of pain and of body, Good also discusses an *inability* to objectify pain (e.g., give it a name). "Even though it occurs within oneself, it is at once identified as 'not oneself,' 'not me'" (Scarry 1985: 52, quoted in Good 1992: 39).
5 An anonymous reader objected to my use of "Western culture" in this fashion, complaining that I cannot speak of one standard in that culture, let alone of unanimity among that culture's members. I agree that I am constructing "Western culture," and that it does not exist in any empirical sense – building on Said (1978), I freely confess to engaging in occidentalizing practice (see Carrier 1992). My use of this phrase is intended simply to contrast "our" (those of us who accept a scientific, materialist vision of the world) sharp distinction between psyche and soma, as compared to various other philosophical systems, some of

which, admittedly, are found in Western nations. Most of the time the pain sufferers at the CPC see mind and body as dichotomous, and they are not unusual in this respect.

6 One patient spoke of a machine newly available in Worcester, MA, that measured pain.

7 This statement, at first sight contradictory, can be made meaningful if we accept that in daily life we seldom become objects to ourselves (see Csordas 1990: 8; also Leder 1990). The patient begins with the preobjective and prereflective experience of his body and finds there is now less "felt pain." Yet when he objectifies himself he finds that there is in fact no decrease in pain. Although patients find it difficult to express this idea, I recorded many other similar remarks – for instance, "You don't dwell on it," i.e., paying attention to the pain or being distracted from it can result in the experience of pain being amplified or dampened. Intention and control of consciousness – Schutz's notion of "working" – are involved to some extent. While one cannot eliminate pain, at least very few graduates of the CPC claim to have done so, one can achieve "less felt pain." Fakirs, of course, report achieving such control over consciousness that they eliminate pain; a Western example is post-hypnotic suggestion.

8 In saying this I am contradicting Schutz himself, who says: "A pain, for instance is not generally called behavior ... But the *attitudes* I assume ... *are* called behavior. I may fight the pain, suppress it, or abandon myself to it ... Experiences of the first type are merely undergone or suffered. They are characterized by passivity. Experiences of the second type consist of attitudes taken to experiences of the first type" (1967: 54, quoted in Bernstein 1976: 143). But is pain *ever* simply "undergone"? No. We would have no "attitude" whatsoever toward a negative stimulus only if we were psychologically pathological, or unconscious, and experiencing pain requires that one be conscious. Schutz is wrong; pain is not analogous to eye blinks or breaths. If one is not aware of pain, as one is not aware of eye blinks, the pain is simply not there. The causes of pain – electrical or mechanical – may still be there, but the pain is not. Thus, pain is *never* "just" sensation and therefore is not, in Schutz's system, ever just experience. He is indeed correct in saying that pain "is not generally called behavior," but I would argue that a phenomenological approach would call pain exactly that, because the very definition of pain implies the presence of what Schutz refers to in the above quote as "attitude." We can conclude, then, that pain is indeed "behavior," the "meaning-endowing experience of consciousness' (Bernstein 1976: 143). In the lived world, pain is always more than sensation and always more than experience.

9 I use "emotion" and "sensation" as opposed to "feeling" because of the (very instructive) mind–body ambiguity of that word: it is possible to have feelings which are strictly sensations or strictly emotions. One author, in fact, defines pain as the "third pathologic emotion" (Swanson 1984), although his conclusions are quite different from those of this paper. On the distinction between sensation and emotion in relation to culture theory, see Jenkins (1994).

10 For an interesting discussion of physical heartache as a synesthesia of several densely packed metaphors see Csordas (1994: 115–23).

11 Merleau-Ponty (1962: 303) describes the body as "a certain setting in relation to the world." This world is constituted of both natural and cultural environments:

"consciousness projects itself into a physical world and has a body, as it projects itself into a cultural world and has its habits; because it cannot be consciousness without playing upon significances given either in the absolute past of nature or in its own personal past" (ibid.: 137).

12 As some apartment dwellers can testify, being forced to experience very loud music would be another example; we might ask, however, is this still the world of lived music or another example of the pain-full world?

13 Which, since the primary meaning of pain always refers to something experienced in the body, I am calling metaphorical pain.

14 In a certain sense, these reported feelings of *communitas* about pain contradict Scarry's (1985: 4) assertion that "pain comes unsharably into our midst as at once that which cannot be denied and that which cannot be confirmed." However, she is absolutely correct that pain cannot be confirmed using ordinary means.

15 That a pain sufferer alone in a room might communicate his pain to an other in his imagination rather than a real other is an important difference, but does not detract from the basic position that pain itself is not made apparent to any but the experiencer unless he or she chooses to engage in a socially motivated, intended action.

16 Scarry (1985: 54) draws attention to negative consequences of communicating pain with her example of newspaper articles containing headlines such as "A pain is a pain if you complain" and "chronic pain can make you one."

17 However complex their relations, the pain-full world does articulate with the everyday world. For example, as noted above, the preobjective, pre-abstract world is nonetheless a world with cultural meaning, the meaning provided in part from the everyday world of "working" (Schutz 1971: 212–18). A non-pain example of a crossover between worlds is Oliver Sacks' (1984) description of "musicking" himself down a cowpath after an injury.

18 It is also true that I collected many, many comments from patients saying that for the first time they feel they *can* talk about their pain. Clearly, at the CPC rules exist which both encourage and discourage talk about pain; it depends upon context.

19 All patients do agree that it is easier to talk about pain at the CPC than on the outside. Note that the difficulty is in talking about *pain* – patients may or may not have difficulty talking about other problems they have.

20 Language about pain in the evening workshops does sometimes come from a third discourse of alternative medicine. For example a session on "love therapy" discusses Bernie Siegel (1986), Kaufman (1976) and Matthews-Simonton et al. (1978).

21 Note that I am including individuals who suffer intractable pain from well-known causes here. Although in one sense they already have a "name" – a diagnosis of arthritis, for instance – they still search the lexicon for a word that describes their condition more comprehensively; the word that will account for why they, unlike some other arthritis sufferers, suffer *so much*.

22 "Metaphorical pain" is not a totally felicitous choice because it carries a connotation of unreality. "Suffering" is a possible substitute if it is made clear that "suffering" is complementary to "pain" in this context (cf. Kleinman and Kleinman 1991).

REFERENCES

Austin, J. L. (1962) *How to Do Things with Words*. Cambridge, MA: Harvard University Press.

Bernstein, Richard J. (1976) *The Restructuring of Social and Political Theory*. Philadelphia: University of Pennsylvania Press.

Black, R. G. (1979) Evaluation of the Complaint of Pain. *Bulletin of the Los Angeles Neurological Society* 44 1–4: 32–44.

Bokan, J. A., R. K. Ries, and W. J. Katon (1981) Tertiary Gain and Chronic Pain. *Pain* 10: 331–5.

Bonica, J. J. (1976) Organization and Function of a Multidisciplinary Pain Clinic. In M. Weisenberg and B. Turskey, eds., *Pain: New Perspectives in Therapy and Research*. New York: Plenum Press, pp. 11–20.

Bourdieu, Pierre (1977) *Outline of a Theory of Practice*. Richard Nice, trans. Cambridge: Cambridge University Press.

Carrier, James (1992) Occidentalism: The World Turned Upside-Down. *American Ethnologist* 19 (2): 195–212.

Corbett, Kitty (1986) Adding Insult to Injury: Cultural Dimensions of Frustration in the Management of Chronic Back Pain. Ph.D. Thesis, University of California, Berkeley.

Csordas, Thomas J. (1990) Embodiment as a Paradigm for Anthropology. *Ethos* 18: 5–47.

(1993) Somatic Modes of Attention. *Cultural Anthropology* 8: 135–56.

(1994) *The Sacred Self: A Cultural Phenomenology of Charismatic Healing*. Berkeley: University of California Press.

Csordas, Thomas J. and Jack A. Clark (1992) Ends of the Line: Diversity among Chronic Pain Centers. *Social Science and Medicine* 34(4): 383–94.

Good, Byron J., (1992) A Body in Pain – The Making of a World of Chronic Pain. In M. J. Good, P. Brodwin, B. Good, and A. Kleinman, eds., *Pain as Human Experience: An Anthropological Perspective*. Berkeley: University of California Press, pp. 29–48.

Hilbert, Richard A. (1984) The Acultural Dimensions of Chronic Pain: Flawed Reality Construction and the Problem of Meaning. *Social Problems* 3(4): 364–78.

Jackson, Jean (1991) The Rashomon Solution to Chronic Pain. Paper given in session on Narrative Representation of Experience: Stories of Illness and Healing. American Anthropological Association Annual Meetings, New Orleans.

(1992) "After a While No One Believes You:" Real and Unreal Chronic Pain. In M. J. Good, P. Brodwin, B. Good and A. Kleinman eds., *Pain as Human Experience: An Anthropological Perspective*. Berkeley: University of California Press, pp. 138–68.

Jackson, Michael (1989) *Paths Toward a Clearing: Radical Empiricism and Ethnographic Inquiry*. Bloomington IN: Indiana University Press.

Jenkins, Janis H. (1994) Culture, Emotion, and Psychopathology. In Hazel Markus and Shincbu Kitayama, eds. *Culture and Emotion: Multidisciplinary Perspectives*. Washington, D.C.: American Psychological Association Press.

Johnson, Mark (1987) *The Body in the Mind: The Bodily Basis of Meaning, Imagination, and Reason*. Chicago: University of Chicago Press.

Kaufman, Barry N. (1976) *Son-Rise: One Family's Journey from Hopelessness to the Triumph of Love*. New York: Harper and Row.

Kirmayer, Laurence J. (1992) The Body's Insistence on Meaning: Metaphor as Presentation and Representation in Illness Experience. *Medical Anthropology Quarterly* n.s. 6: 323–46.

Kleinman, Arthur and Joan Kleinman (1991) Suffering and its Professional Transformation: Toward an Ethnography of Interpersonal Experience. *Culture, Medicine, and Psychiatry* 15: 275–302.

Leder, Drew (1990) *The Absent Body*. Chicago: University of Chicago Press.

Matthews-Simonton, Stephanie, P. Carl Simonton, James L. Creighton (1978) *Getting Well Again: A Step-by-Step Self-Help Guide to Overcoming Cancer for Patients and Their Families*. New York: Bantam.

Melzack, Ronald (1975) The McGill Pain Questionnaire: Major Properties and Scoring Methods. *Pain* 1: 277–99.

Merleau-Ponty, Maurice (1962) *Phenomenology of Perception*, C. Smith, trans. London: Routledge and Kegan Paul.

Rorty, Richard (1979) *Philosophy and the Mirror of Nature*. Princeton: Princeton University Press.

Rosaldo, Michelle (1984) Toward an Anthropology of Self and Feeling. In R. Shweder and R. LeVine, eds., *Culture Theory: Essays on Mind, Self and Emotion*. Cambridge: Cambridge University Press, pp. 137–57.

Sacks, Oliver (1984) *A Leg to Stand On*. New York: Harper and Row.

Said, Edward W. (1978) *Orientalism*. New York: Pantheon.

Scarry, Elaine (1985) *The Body in Pain: The Making and Unmaking of the World*. New York: Oxford University Press.

Schaeffer, P. (1983) Learning to be a Chronic Pain Patient. *Aches and Pains* 4 (4) (April): 21–5.

Schutz, Alfred (1967) *The Phenomenology of the Social World*, G. Walsh and F. Lehnert, trans. Evanston, IL: Northwestern University Press.

  (1971) On Multiple Realities. In Schutz, *Collected Papers: Vol. 1, The Problem of Social Reality*. The Hague: Martinus Nijhoff.

Siegel, Bernie S. (1986) *Love, Medicine and Miracles: Lessons Learned about Self-Healing from a Surgeon's Experience with Exceptional Patients*. New York: Harper and Row.

Swanson, David W. (1984) Chronic Pain as a Third Pathologic Emotion. *American Journal of Psychiatry* 141 (2): 210–14.

10　The individual in terror

E. Valentine Daniel

> It is pain coming out. Those who see him are terrified. Of course, the boy is also terrified. But that is good. The pain is coming out.
>
> (Rajasinga Vaittiyar[1] on the near-convulsive shivers experienced by a freed victim of torture.)

This chapter is about the terror of torture, experienced and remembered. It is also about beauty and about pain, the attribute of virtual unrepresentability they both share, and their complicities and complexities. And it is about the individual in terror.[2]

Representing the beautiful

In their attempts to account for the aesthetic in culture, three semiotically inclined anthropologists have understood the aesthetic as something located in signs that are difficult, if not impossible, to re-present. For this reason, Roy Wagner (1985) writes of "symbols that stand for themselves"; Steven Feld (1982), of "autoreferentiality"; and Nancy Munn (1986), borrowing from Peirce (1931: 2.244, 248, 254),[3] of "qualisigns" – that is, of signs that are mere qualitative possibilities in contradistinction to signs that are actualized and/or generalized. Even though all three cultural aestheticians then go on to find ways of representing the unrepresentable in apparent contradiction of their opening premises, the point still holds that any representation of the aesthetic is by its very nature incomplete and vague, although richly suggestive, which no number of objectifications can exhaust. In this regard Munn's choice of the Peircean category, "qualisign," is instructive. It behooves us to recall Peirce's claim for signs and signification: to wit, that every sign or representamen is only a sign in potentia unless and until it stands for a second, its object, and to a third, its interpretant, in some respect or capacity (1931: 2.228, 2.274). That is, a sign *qua* sign, must be constituted of three correlations: a representamen, an object and an interpretant. In so far as what Peirce called a "qualisign" is a label for the representamen alone, disregarding for the purpose of analytic attention its object and interpretant, it is not a completed sign but a sign of possibility –

waiting, as it were, for its appropriate object and interpretant to introduce it into significance. Therefore, Nancy Munn's use of "qualisign" to describe the manner in which "the fame of Gawa" is carried forth from one Melanesian island to another is not a "qualisign" in the strict sense but a qualisign that has broken out of its confinement of mere possibility and found its objects to represent and its interpretants to carry forth these representations into semiosis. Nevertheless, Munn is justified in highlighting the qualisignificant aspect of "the fame of Gawa," to the extent that neither its objects nor its interpretants are transparent to native or anthropologist, as for instance an indexical sign such as smoke standing for fire, or a symbolic sign such as a word standing for a thing might be. To put it differently, a qualisign is a sign that admits to the inexhaustibility of its representational mission; one never gets to the bottom of it. As for its significant effect – technically called the interpretant – it is rarely logical but mostly immediate and emotional. This is also the nature of the aesthetic. And therefore, any given representation of beauty even when interpreted as such, only reveals the profundity of the beautiful. Or to put it differently, beauty is best represented by something that represents beauty's unrepresentability. To return to a more technical formulation we may say that a sign of beauty bulks large with qualisignification. We shall soon return to beauty and qualisignification, but only after a brief meditation on beauty's relationship to pain.

Beauty and pain

In the dominant philosophical traditions, or "the great traditions" – to borrow a concept from a Redfieldian anthropology of yesteryear – of the West, the aesthetic has come to be identified with the beautiful, the pleasurable, and the tasteful. This tradition extends from Plato and Aristotle to Hume and Kant, and to a whole range of neo-Kantians. It is, however, the little traditions of the likes of Maurice Blanchot, George Battaile, Walter Benjamin, some of the surrealists, Freudians such as Julia Kristeva, and of course the Sadians, that have revealed to us, if not the complicity of pain in pleasure, then at least the confusion of pain with pleasure. But rather than invoke the little traditions to illustrate the complicity and confusion in question let me briefly go to the great tradition itself.

For a start, one would be hard-pressed to read Kant's famous definition of beauty as "disinterested delight" without experiencing a twinge of recognition of the torturer brought to our attention by Elaine Scarry and others. Or how can one read Schopenhauer who, extending Kant's definition, locates the essence of the aesthetic in "the liberation of the will," and not also see the torture victim who has been liberated from his will? Look furthermore at the mirroring of yet another one of Kant's definitions of beauty in the light of what we know about pain in the context of torture and interrogation. "Beauty is," Kant (1951 (1799): 17.236) says, "the form of the purposiveness of an object so far as this is perceived in it without any representation of a purpose." To have a purpose is to be suited for some use. Here we seem to have an instance in which we say that something is purposive but with no identifiable purpose. But note that Kant does not locate beauty in the purposiveness of an object but in what he calls the *form* of that purposiveness. Ted Cohen (1982: 231) parses this definition as follows:

Because it is purposive in this way: it does suit some use, namely, the use I make of it ... The sense that it makes consists entirely in its making sense to me. Its purposiveness consists in my finding it purposive. My finding it purposive (which is a finding realized as my feeling of pleasure) takes the place of the genuine purpose that in all other cases accompanies purposiveness. When I engage the object something occurs which Kant calls the harmony of my faculties of cognition. But this harmony (my sense that the object makes sense, is coherent) is achieved without the use of concepts, and without a concept I cannot be regarding the object in terms of any determinate purpose. And yet I find it purposive (for my faculties of cognition). What I encounter is mere sense, mere suitability. Mere suitability of x without any definite y for which x is suitable for, is suitability without material context: x displays only the *form* of purposiveness.

Where is the analog in this to pain in the context of "interrogational" torture? With the appropriate substitutions, Kant's definition would read something like this: "Pain is the form of the informativeness of a victim so far as this is perceived in him without any representation of information." As Scarry, Shue and other students of torture have remarked, the vast majority of those on whom the pain of torture is inflicted because of their purported informativeness have nothing to inform. Shue, who proposes the distinction between interrogational torture and terroristic torture, goes on to admit that "it is hardly necessary to point out that very few actual instances of torture are likely to fall entirely within the category of interrogational torture" (1978: 134). To continue with the appropriate substitutions in the quote from Cohen,

because pain is informative in this way: it does suit some use, namely, the use the torturer makes of it ... The sense that pain makes consists entirely in its making sense to the torturer. Its informativeness consists in the torturer finding it informative. The torturer's finding it informative (which is a finding realized as a feeling of pleasure) takes the place of the genuine information that in all other cases accompanies informativeness.

"Pleasure" may not be exactly what the torturer is likely to admit to feeling. The three torturers[4] whose accounts I had the dubious privilege of hearing, lay claim to a quest for the truth. We may call this the pleasure in "truth." In the words of one of the ex-torturers I interviewed, "you have got to beat them [a generic expression for torture] in such a way until they tell you exactly what happened: no more, no less. Then you know that the beatings [pain] were *just right*" (italics, mine). This excerpt from my transcription of my fieldnotes merits reading alongside Alberti's celebrated definition of beauty as "a harmony of all the Parts, in whatever Subject it appears, fitted together with such Proportion and Connection, that nothing could be added, diminished or altered, but for the Worse" (1966 [1485]: 113). All three torturers interviewed were hard-pressed to come up with a specific instance of "truth" worthy of their efforts. The circumstances of the "interview-conversation" were such that there seemed to be no apparent need for them to conceal any "truthful" information they might have obtained. Their floundering and groping around for something persuasive to offer me when asked seemed to indicate that they were clearly not sitting on anything meriting the label, "top secret." Nevertheless, they continued to insist that those who were tortured were deserving embodiments of pain. This leads me to continue with my adaptive appropriation of the quote from Cohen,

When the torturer engages the victim something occurs which we may characterize as the harmony of the torturer's faculties of cognition. But this harmony (the torturer's sense that the victim in pain makes sense, is coherent) is achieved without the use of concepts, and without a concept the torturer cannot be regarding the object in terms of any determinate information. And yet the torturer finds the victim-in-pain informative (for the torturer's faculties of cognition). What s/he encounters is mere sense, mere suitability. Mere suitability of x without any definite y for which x is suitable, is suitability without material content: x displays only the *form of informativeness.*

The knowledge of the connection between beauty and pain in "the little tradition," may be commonplace. My detour through the great tradition rather than summoning this little tradition is intended to suggest that there is perhaps a stronger case to be made that the sensory mode called the aesthetic ought not to be limited to the beautiful but must be ready to admit a most unwelcome member into its domain, pain. Recalling that signification by definition entails a three-place-predication of representamen, object and interpretant, if virtual unsignifiability is what characterizes what we have called the qualisignification of beauty, then what are we to do with the virtual unsignifiability of pain? In short, if unsignifiability (unrepresentability and uninterpretability) were the criterion, then this member's claim to regnancy in the aesthetic domain would be certainly assured, especially in its sheer resistance to language. In her book *The Body in Pain*, Elaine Scarry (1985: 4) says of the relationship of pain to language: "Physical pain does not simply resist language but actively destroys it, bringing about an immediate reversion to a state anterior to language, to the sounds and cries a human being makes before language is learned." In so far as the aesthetic is concerned with the sensory, or rather with throbbing percepts, percepts that push against the conceptual membrane that encloses the world of active semiosis and articulate speech, pain is one with beauty.

There is however, one – at least one – difference. While beauty too puts language on trial it does so in a manner quite different from the way pain does. Beauty finds language wanting because of beauty's profound inexhaustibility; pain finds language wanting in pain's excruciating particularity. If beauty as a qualisign is pregnant with inexhaustible possibilities, pain is a *sinsign*[5] that is exhausted in its simplicity and singularity.

If, as I have indicated, a qualisign is construed to be a pre-sign or a hyposign that contains – in its double sense – significance, its freedom lies in its potential for expansion in semiosis, in becoming more than a mere qualisign; it "lies" in locating its object for which it stands and an interpretive manner in which it stands. And here I intend the pun in "lie": its situatedness and its mendacity. In finding itself a place or position it loses a measure of its truth, its integrity. Be that as it may, the manners in which beauty on the one hand and pain on the other find their freedom are different. The principal mode of signification in which beauty finds its expression is in iconic signs[6] – in metaphors and in objects that partake of beauty's qualities in different measures.[7] This objectification is generous. It opens out to the world, inviting further signs, objects, and interpretants to partake of its bounty, of its essence. Pain, by contrast, when embodied, closes in on itself. Where beauty extends itself, pain finds affirmation in its intensification. Beauty repressed can be painful, pain expressed is susceptible to incredulity.

Pain is highly localized. Its outermost limit is the boundary of the victim's body. Its inner limit can be as small as a point in one's foot where a nail is being pounded in. And no one pain is like any other. Two examples should help us understand these limits.

The individual in terror

In late November 1983, Benedict was rounded up with twenty other young men by the army in Batticaloa in eastern Sri Lanka. He claimed he was innocent then and he claims he is innocent now, and had assiduously avoided the militants operating in the area. He was interrogated by a sergeant, then by a captain, and then by a man in civilian clothes, then by one in uniform, then by a corporal, and then by another captain, and soon he lost count of who and how many had interrogated him. They did not believe him. There was another boy who, under torture, named him as one of the militants. Benedict was tortured by the soldiers repeatedly in order to extract "information" from him. He was beaten with two-foot long plastic pipes filled with sand. Beatings with sand-filled pipes have an unusual capacity to cause pain without breaking the skin as easily as do other implements. One does cough up blood from the beatings on one's chest and back with one of these implements, but skilled torturers see to it that the skin does not break. The heaviest beatings are inflicted on the soles of feet, even to where the skin actually breaks. But as for the rest of the body, fluid builds up under the skin. The whole body swells up. The face is spared, and above all, care is taken not to break the skin. This is evidence, presentable to the investigating magistrate, that information was not extracted under torture. The victim's greatest certainty, his pain, is paradoxically also the magistrate's locus of doubt. Pain stops at the skin's limit. It is not shareable. Benedict did not give the torturer any "information."

From the local army camp he was transferred to the now-infamous Boosa camp in the island's south. In this camp he was handed over to the expert care of the dreaded Special Task Force. There, too, there were numerous torturers; the worst among them was a young woman in uniform. Of the great variety of highly specialized torture techniques he was subjected to, he recalls the swallowing-the-saliva torture as the worst. In this technique one is made to swallow one's own spittle, once every six seconds, for as many as 200 times, before a fifteen-minute break. This causes excruciating pain in one's insides: starting with the throat, and head, extending through the gullet to the chest and stomach. If he stopped swallowing before it was time, he was beaten with a sand-filled plastic pipe, he was burned with a cigarette on his feet or where the rope handcuffs had already cut through the skin, or a live wire was held to his genitals. "Suddenly one pains all over, inside and out." Finally he was given the option of not providing information but answering "yes" or "no" to a series of questions. A kind of multiple-choice over essay questions.

In 1987, when I interviewed Benedict in Colombo soon after his release, he bore no manifest evidence of having been tortured except for black spots on his brown wrist, which he saw but I could not distinguish from the black marks left by the rope handcuffs. He showed me his feet. He could see the stripes caused by the beatings and the burn marks caused by cigarette butts. I couldn't see these, either. Either they had healed without leaving behind any traces, his feet didn't scar, or he was never tortured but claims that he was. The last was suggested to me by the cook who happened to walk in on my interview. He, too, was a Tamil. But he was not the only one to doubt. Benedict himself doubted the accounts of other torture victims. Nevertheless, three months after his release, he told me:

Even now, some times, like the other day when the BBC man was interviewing me, my body feels like a bubbling candle, filled with pain, inside and out, all over the skin and all over inside. One pain.

My second example is based on a torture victim I saw in 1984, in one of the Tamil militant training camps in southern India. The torturer this time was not a Sinhala government soldier or one of the white men from the Keeny Meeny Services, an outfit of British mercenaries, who, it was alleged, trained members of the Sri Lankan Special Task Force. Rather, it was a Tamil; and so was the victim.

As late as May 1983, the sum total of Tamil militants in the Jaffna peninsula numbered no more than 300, armed with largely primitive weapons and with virtually no citizen of Jaffna willing to rent his or her home to these ragtag fighters. The anti-Tamil riots of July of that year resulted in massive voluntary enrollments, and recruiters from the five main militant groups were competing in a field wherein the harvest was plentiful. Most enlistees did not know which group they were enlisting in and what or whether it made any difference. All they knew was that they would be shipped off to a camp somewhere, most probably in India, where they would be trained in armed battle and return to avenge their kinfolk, friends, and fellow Tamils who had lost their lives and possessions in the riots, and help establish the independent state of Eelam.

As it turned out, the "training" in the "training camps" was limited to a week's weapons training and months of waiting in a second camp where calisthenics, malnutrition, indoctrination, and despotic intimidation by a twenty-one-year-old camp commander was the daily fare until and unless orders came "from above" that there was a mission to be carried out. At times the commander's intimidation tactics turned to forms of bodily torture. Unlike Boosa, the torturer rarely, if ever, professed the pretense of "information gathering." It almost always began as a form of amusement for the commander and a few of his lieutenants. Amusement would switch into fantastic illusions of power on the part of this commander without warning. The more the victim cowered and shrank in pain, the more the commander expanded his swelling power. And what began as amusement would often conclude with the pretense of it having been an attempt at information gathering or punishment for some vague infraction. No two individuals were allowed to carry on a conversation for however brief a moment without a third person being present. According to one informant, "I am sure most of those in there wanted to escape. But there were a few who were the commander's men. The thing is no one knew who was and who was not." However, from my own interviews with ex-inmates of this camp, it was hard to say whether boredom, disenchantment with "comradism," or the commander's sadism was the most unbearable in this camp.

There were several daring attempts at escape. If caught, an escapee would face one of two forms of punishment. The first and the worst would be execution after being accused of being a traitor to the movement, or a spy for one of the rival militant groups or for the Sri Lankan government. The lesser punishment would be to place the victim in solitary confinement for thirty days.

One of the residents of the camps, Guhanesan by pseudonym, had been recruited from the Mannar District at a time when his wife was seven months pregnant. On the 270th day after Guhanesan's arrival in the camp, a new recruit arrived from Mannar. He brought Guhanesan the news that his wife had given birth to a son. Guhanesan, a somewhat senior camp member by then, asked the camp commander's permission to make a quick boat trip to see his wife and son, promising to return to camp within seven days. The young commander instantly refused his permission and, furthermore, accused him of harboring "bourgeois-feelings." A much-touted part of the movement's manifesto was that forsaking family was the true mark of a fighter. In the words of my main informant on matters of this camp, "everybody knew that this was one rule not to be taken seriously, because no Tamil fighter ever forgot his family; not even the camp commander who visited Tiruchirappli or Madras whenever members of his family were passing through these cities." Guhanesan defied the commander's orders and with the help of some of his comrades escaped from the camp. However, as promised, he returned on the seventh day, even though rough seas and the stepped-up surveillance by the Sri Lankan army in the forests in the northwest region of the island allowed him only six short night-hours to spend with his wife and son. He returned to the camp in order to prove to everyone that he was not a deserter, and was willing to undergo the thirty days' solitary confinement, if necessary.

Instead of solitary confinement, the camp commander personally tortured him by hammering a nail up the sole of his right foot and left it there overnight. The next day, the nail was removed from the swollen (and presumably infected) foot, and then he was beaten on his back with a heavy baton. "And then," in the words of my informants, "his legs were broken, as a warning to the others of what could happen if orders were disobeyed." In mid-1984, Guhanesan, paralyzed below the hip, was dragging himself around the camp grounds on a tire like a beggar. Orders were that he be described as a beggar to outsiders, but to the residents of the camp, he was to serve as a "living example." He also contributed to the humor of the other militants-in-waiting. Even though he was completely paralyzed below the waist, and could not feel even a live cigarette butt on his foot, he groaned in pain (in "imaginary pain," according to the others), complaining of a nail stuck in his right foot. The excruciating pain was in only this single spot, no larger in circumference than the point of the nail. All attempts by others at locating this spot by feeling and pressing the entire surface of his foot would yield no results. He couldn't feel a thing except for this single, unidentifiable, excruciating spot of pain. This was a contradiction and an impossibility that his more fortunate camp-mates found hilarious.

I have introduced you to two examples in order to illustrate the limits, the particularity, the unshareability, and the incommunicability of pain in torture. There are tens of other instances that I could employ to make the same points regarding pain in torture. The phenomena described are no different from the countless accounts one may find in reports on torture in other parts of the world obtainable from Amnesty International and other human rights abuses monitoring organizations. At this level of experiencing pain it appears that one is unlikely to find any significant effect of culture.

There are, of course, the much more socialized pains: the headache, the toothache, the earache, the back spasm, and the countless other pains that have been given names of recognition in the folklore and diagnostic labels in the medical lore. Their representations are public, available for more than one person to map his or her private experience onto, even though no one pain is like another. And from there sympathy and empathy take over, making the pain in question more or less shareable.

In torture, *Homo individuum* reaches its extreme. Pain's privatization becomes absolute. Its *hic et nunc* is complete. Indeed, I encountered time and time again torture victims who had been subjected to the same tortures by the same torturers in the same camps and jails, and even at the same time, and who – when they finally were capable of speaking about their experiences – denied that their fellow inmates were tortured and accused them of lying.

Torture perverts all dialogic. For the victim's interlocutor is not that person's parent, friend, or child. The interlocutor is not even a mere stranger, but rather the torturer, for whom pain is not a sign of pain, but an insignia of power. And even after the torture victim is set free, regardless of the readiness of friends and kinfolk who had remained free to extend love, understanding, and sympathy, the victim persists in denying the shareability of the individuated pain, contained within the bounds of a private body that is reticent to express itself willingly through any signs in general and incapable of expressing itself in signs of language in particular.

Of course, the newly freed victims did talk. In interview after interview I found these individuals willing to talk about their capture, their cells, their meals, their fellow-prisoners, the weave of their mats and the state of their clothes, food and furniture. Of these they spoke even with some animation, expecting you to be interested. All these details were believable and therefore understandable. But of torture they volunteered very little.[8] Then gradually – with coaxing, with the purpose of the interview undeniably clear – they began to talk. But the interest that animated the recall of the mundane details of prison life suddenly abandoned the intonation of speech. Of course, brutalities were wearisomely enumerated and details enunciated, first with bored parsimoniousness, then in flat-toned recitations devoid of conviction, pausing now and then as if only to wonder how these details could interest anyone. Cruelties that had lain outside speech and outside reason were recited with such syntactic correctness as to become a routine, a mannerism, a set of clichés. I don't believe that most of the victims I interviewed were unwilling to talk about their experiences because it was too painful to do so. There were no signs of contained passion. Rather, attempts to extract information were met with expressions of utter listlessness. I was later to find out that it was not so much boredom that weighed down on the victim but the overwhelming sense of the sheer worthlessness of all attempts to communicate something that was so radically individuated and rendered unshareable. As Scarry (1985: 32) describes it, "the narration as a whole has the quality of a sketch: the experience it describes is utterly clear in its outline but all the emotional edges have been eliminated." In court, this disposition of the victim is exploited by the torturers to make lies of the victim's claims of having been tortured. The passionless listings of atrocities committed by the torturer leaves the judge and the court unconvinced. The unbelievable is also unexplainable and the unexplainable the inexpressible.

Thus, it would be fair to say that pain does not find its way out of its arrested, insignificant state, even when it finds objectification in verbal recitations or in listings of painful experiences. In order for pain to find its freedom into the semiosis of culture, it must pass through the liminal phases of either terror, beauty, or both.

Of course, it is terror (and pain) that hyper-individuate the victim in the first place. But the terror I speak of is a second, therapeutic terror, a seismic after-shock. This terror can take many forms. Of the torture victims I came to know whose pain passed into this terror, terror manifested itself in a variety of ways: in uncontrollable sobbing, in rage, in violent shakings of the body, in visible efforts at restructuring a narrative with conscious self-corrections introduced regarding details and their proper sequences – corrections that even while appearing to be far more confusing to the interviewer than the mechanical recitation of the pre-terrorized torture victim, carried far greater conviction.

In all its manifestations, therapeutic terror begins the process of the de-individuation of the person, the process of disarticulating a hyper-individualized self that had been compacted in pain. Whether or not the victim's pain is reduced in and through terror, I cannot say. I had no means of measuring pain. The Vaittiyar whose words form the epigram of this essay described this terror as "the fear and trembling that comes through remembering terrible acts." In the original Tamil – "*kodurattai ninathu varum payangaramum nadunadukkamum*" – the description of redemptive terror is made up of a reduplicating consonant and vowel series of flaps and trills that onomatopoetically capture and echo the sense of the experience itself,[9] metonymically triggering a veritable shudder of sound-images. *Koduram, nadunadukkam*, and *payangaram* are all offered as translations of "terror" in the Madras University English–Tamil Dictionary (Chattiar 1963). The terror in question – remembered or re-lived terror – dislodges pain from its fixed site. I am suggesting that terror shows pain a way out of its static particularity into a domain of inexhaustible virtuality, a domain that is home to beauty as well. But what we do know for certain is that once pain extends into terror, it becomes more understandable to those with whom the victim returns to live, and reflexively, pain becomes more understandable to victims themselves.

The mother in whose arms her twenty-four year old son is finally able to cry comforts herself as much as her son with the words, "son, you have returned at last." Not, "you are healed at least," but "you have returned at last." Healing, she realizes, will take "as long as it takes to forget what happened." Rajasinga Vaittiyar, the Ayurvedic physician about whom I have written elsewhere, considers body tremors to be good. "It is pain coming out," he says. "Those who see him will be terrified. The boy is also terrified. But that is good, for the pain is coming out."

Pain also finds its exit into semiosis through works of art. Twenty years ago, the Jaffna peninsula, which today is the center of the Tamil separatist movement, was an extremely unmusical and unpoetic corner of the island. To be sure, girls studied the piano for their various certificates in music from the two London-based Schools of Music, the Trinity and the Royal College of Music. Every afternoon, in certain villages of Jaffna, among the Christians, one could hear excruciatingly mechanical and earless attempts at thumping out Bach Fugues or Preludes on untuned pianos. It gave them "culture" and a discount on their dowries. There were also those who learned Karnatic music. The quality of this music as a whole was much better but pathetic in comparison to what one would find amongst a comparable class of people in southern India. Students who went into the humanities (or the "arts," as it is still called) in high school and in the university were treated as intellectual rejects. Teachers of Tamil literature were most frequent targets of jokes. Jaffna liked to think of itself as training sober professionals; physicians, engineers, mathematicians, scientists, and lawyer-politicians. The available poetry and prose were scanty and mediocre.

But now, in the middle of the civil war, street theater is flourishing, and poems that make one tremble are written daily. It is true that on one hand the people of Jaffna as a whole were experiencing an ever-shrinking perimeter of freedom as if it were a collective body in pain, lost to the outer world as much as the outer world was lost to it. On the other hand, as if to regain the lost world, the voice, especially in its aesthetic mode, provides the needed self-extension as a means of occupying a much larger space than the pain-inflicting constraints of the military presence allowed it. The poem below is one written after the Indian Peace Keeping Forces landed in Jaffna.

Ever since the civil war began, throughout the peninsula, there were blackboards fixed to trees near certain well-known intersections. On these boards, every morning, the Tigers wrote the names of those among them who had been killed. Every day for four years, the boards unfailingly had fresh names on them. But on the day following the Indian Peace Keeping Force's arrival, the blackboards had been wiped clean and stayed clean. And so the poem goes:

> otinal aval patunki
> antru entrum pol-otunki
> parttal kan kavarntu
> mantar per attanayum parttum
> parkka viriyil terittate
> verumpalakai
> mutal veruntal
> pin amarntal
> ematramataintal
> tuyarankontal
> natu natunkinal
> paniyunta alamum
> tannai maranta notiyil
> natunki nimirntatu.

> On that day,
> unlike on all other days
> when having sneaked to the tree
> to hide and read with longing
> the names of the dead,
> she froze as if smitten
> by the markless board,
> and settled,
> haunches on heels,
> first disappointed,
> then saddened,
> to shiver and shiver again
> while the great dew drunk banyan tree too,
> in a moment's self-forgetfulness,
> shook
> and then stood still.

The poem contains within it both pain and, in its shivers, the ingredient of redeeming terror.

In my limited fieldwork I did not find any works of art, in poetry, prose, painting, or sculpture, wrought by a victim of torture. This is not to say that such artists do not exist. However, I have observed victims moving from a condition of speechlessness or "autistic speech" to the ability to form elementary metaphors. The appearance of the latter is the earliest sign of pain's entry into cultural semiosis. This process is marked by pain's retreat from its sinsignificant state into pure qualisignification and its re-emergence and re-expression through objectification in iconic metaphors. Consider the following excerpts from two separate interviews I had with Benedict, in June 1987 and in September of that same year.

Interview 1

(After lengthy preliminary introductions, and assurances from those who were responsible for his release – therefore those whom he could trust without reservations – that I was alright and that he should tell me everything.)

DANIEL: How long were you in Boosa?
BENEDICT: Twenty-four months.
DANIEL: Before then?
BENEDICT: In Batticaloa.
DANIEL: Where in Batticaloa?
BENEDICT: In the police jail.
DANIEL: Tell me a little of what happened.
BENEDICT: [After three minutes' silence.] What do you want [to know]?
DANIEL: Why did the police arrest you?
BENEDICT: [After a further lengthy silence.] The police didn't, the army did.
DANIEL: And then?
BENEDICT: And then what [rhetorically]? Then all that happened happened.
DANIEL: Is it difficult for you to talk about what happened?
BENEDICT: What difficulty [lightly]?
DANIEL: Is there any reason why you don't wish to talk about anything? I don't wish to force you to talk.
BENEDICT: There is nothing like that. [Silence.]
DANIEL: Tell me, what were you doing when the army came and arrested you?
BENEDICT: I was in the shop.

This preliminary bit of dialogue took almost fifteen minutes. After about half an hour more of this pace, we broke for lunch. We continued the interview that afternoon.

DANIEL: Tell me about Boosa.
BENEDICT: What do you want me to tell you?

DANIEL: Oh. What time you got up, when you ate, what you did all day, what time you went to bed. Things like that.

BENEDICT: When they tortured us we would be awakened at different times.

DANIEL: How did they wake you up?

BENEDICT: They poured water on us. They hit us with a stick.

DANIEL: So your clothes and bed would be all wet.

BENEDICT: There was no bed. A grass mat with a single weave, sometimes. Mostly it was the floor. The floor was always wet. There was mud. The room was only so big. At different times different numbers of people would be in it. The first month there were four.

DANIEL: The same four people.

BENEDICT: Never the same four.

[A later segment of the same interview]

BENEDICT: We had to run on our heels so that at all times only one foot was on the ground. While we ran soldiers would hit us on the calves with PVC [polyvinyl chloride] pipes.

DANIEL: You said something about cigarettes.

BENEDICT: Yes they burned us.

DANIEL: Often?

BENEDICT: Sometimes often. Sometimes not so often.

DANIEL: Where did they burn you?

BENEDICT: Here. Here. Here. Everywhere.

DANIEL: Not on the face?

BENEDICT: Not on the face.

DANIEL: What did they do to the wounds?

BENEDICT: Nothing. Sometimes they put chilies in them.

DANIEL: Did they ever give you first aid?

BENEDICT: Yes.

DANIEL: How often?

BENEDICT: Depends.

DANIEL: Depends on what?

BENEDICT: I don't know ... Sometimes we would be allowed to do exercises. We always tried to exercise in our cells. The last year it became crowded. Then we would talk. The EPRLF[10] chap used to teach us about Marx and the revolution.

DANIEL: Did you learn about Marx?

BENEDICT: It was a diversion.

DANIEL: Was anyone from the other [militant] movements with you?

BENEDICT: Everybody. Militants, non-militants, informers, older men, young boys ... about twelve.

DANIEL: Did these other groups try to teach you, too?

BENEDICT: Not so much. Some EROS people also did.

DANIEL: The Tigers?

BENEDICT: They did not say much.

DANIEL: But there were Tigers [LTTE]?

BENEDICT: They did not say that they were Tigers. But they did not say much. Tigers don't talk much.

DANIEL: TELO?

BENEDICT: There were some of them, too. Who knows?

DANIEL: Were there fights and arguments among the groups in the camp?

BENEDICT: No. All were happy when soldiers were killed. But then, whenever soldiers were killed we would be tortured.

DANIEL: Are you going to join a group?

BENEDICT: I don't know.

DANIEL: Which group do you think you'd join if you chose to join a group?

BENEDICT: Whatever you say.

DANIEL: But [laughing] I am not recruiting for any group.

BENEDICT: Or X [the person responsible for his release] says.

DANIEL: What does X say?

BENEDICT: I don't know.

DANIEL: Was Sudharshan also tortured?

BENEDICT: I don't know.

DANIEL: But you were in the same camp.

BENEDICT: They all say that they were tortured.

DANIEL: But Sudharshan showed me his scars.

BENEDICT: Lies.

DANIEL: How about Nathan?

BENEDICT: All lies.

DANIEL: But he too showed me scars. He swore.

BENEDICT: You can get scars from rats biting.

DANIEL: Rats?

BENEDICT: Yes, there were rats sometimes. Here. Look [showing his heels].

DANIEL: But now why do you think Nathan, Sudharshan, and all the others would have sworn to me had they not been tortured?

BENEDICT: The fellow who told the army that I was a militant also swore.

DANIEL: So you don't think anyone else was tortured.

BENEDICT: They might have been. I heard them scream. But who knows if it was for real. Sometimes they have tape recorders, too [i.e. tape-recorded screams]. All I know was they tortured me.

DANIEL: But you said that others were tortured with you sometimes in the same room.

BENEDICT: I don't know. What if they were acting?

In the last part of the interview the paradox of pain – one man's certainty providing the other man's primary model of what it is to doubt – is brought to a pathetic pitch. One even doubts one's fellow-sufferer's pain.

Interview 2 (discontinuous quotes)

Imagine having to eat where you also have to pass wastes.

This woman torturer was the worst. If you just looked at her, she could be your friend's sister. Young. I think she was a Muslim. But when she opened her mouth. Father! I can't even repeat her words. Bad words. Like a sewer (narakal)

You couldn't find a handful of rice without stones in it. [We] just touched it like that and pushed it aside. You wouldn't even feed your dog rice like that. There was this thing called soti [a watery gravy] they served us. It smelled of urine. They used to laugh when they served us that.

This one fellow, they beat the bottom of his feet in such a way that the skin from one sole just fell off. Just like that. Like a tire. You have seen these pieces on the road. Like that. And underneath, skin like a baby's.

And finally,

Even now, sometimes, like the other day when the BBC man was interviewing me, my body feels like a bubbling candle, filled with pain, inside and out, all over the skin and all over inside.

I shall have more to say about this last quote. But for now it is sufficient to observe two features of the interview: the invoking of the domestic and the emergence of metaphors and similes: sewer-like mouth, tire-like skin, and candle-like body.

In these quotes we hear Benedict finding it reasonable to have expected something more like what was true in his home or any civilized home. Eating where one passes bodily waste is unthinkable at home. The outhouse or bathroom is as far removed as possible from the kitchen in a Tamil home. Stones in rice is an acute embarrassment to the women of the home and is often enough for the husband or father to stomp in anger or to pick a fight or to berate the wife. It is only reasonable for a domesticated man, a civilized man, to expect stone-free rice. A dog, a domestic animal, is brought in as an illustration. It is not any dog, but "your dog." Perhaps the primordial social unit wrenched away in prison and under torture was the reality of belonging to a home and a family. Indeed, kinship is referred to in the second quote. The torturer is likened, at first glance, to a "friend's sister." In earlier interviews when and if references to home and kin were made they performed a strictly referential function, as part of a minimally descriptive account. They were never employed in any figurative or extended sense.

Finally, I wish to consider the only torture victim I met who came from among the plantation Tamils. There weren't many of them in the camps, but Murugesu was arrested for being in the wrong place at the wrong time. One thing led to another; he was accused, tortured, made to confess, tried, found innocent, and released. The whole process took one year.

Murugesu was much more forthcoming than the others I had interviewed. And in my judgment I thought he was adjusting quite well to his past experiences and his present options. But Murugesu's mother did not think so. I tried to find out what exactly she thought was wrong with her son. All she said was, "He isn't like before." I couldn't get any more details than that, since I had not known what he had been like. His father, a retired and sickly plantation laborer, only partly shared his wife's concerns. But the *kotanki* (the local diviner-healer-priest) seemed to understand the mother's concerns quite clearly. His diagnosis was that Murugesu had a demon in him, a demon that had entered him when he was in Boosa camp. And there was no point in trying to exorcise this demon because he was all tied up inside with rope and was asleep. The *kotanki*'s task was to wake up the demon and make him tremble. This was to be done by offering him blood sacrifices and by singing mantras to the accompaniment of the *utukku*-drum. "In India," he said, "they use a horsetail-whip to beat the demon into waking up and trembling. My grandfather used to do that. But nobody here knows how." I asked him how big the demon was. He said that it was big as his fist, "like a swollen boil." Without any further evidence, I hypothesize that this bound, dormant demon is pain that has to be terrorized into wakefulness before being made to move out into the world.

During the preparations for the ritual, Murugesu's father asked me if I was a Christian. I mumbled an answer and he asked no further, but moved on to tell me tales of life on the plantation during his youth. One sight he said he could never forget. It was that of the Anglican catechist, a dark man from the Nadar caste, walking up the hill with his violin in one hand and a magic lantern in the other and a shoulder bag with a Bible in it. And of all the pictures he showed, the one that they all liked the most, and had asked him to show them over and over again, on every visit, was that of Christ dying on the cross. "How beautiful! What pain! It gave us peace" he said. *(Evvalavu vadivu! Enna vetanai! Namakko oru camatanam)*

The next day, Murugesu's father asked me if I could take them to the Roman Catholic church in the town nearby. I told him that I didn't quite know how to worship in a Catholic church but that I would like to accompany them. So we went to St. Mary's Church. The mass was taking place. I, knowing only slightly more about what to do in a Catholic church than did my friends, walked up to the first pew and knelt down alongside a few other worshipers. Both Murugesu and his father prayed, with eyes fixed on the icon of Jesus on the cross. Between us and the priest and the altar and the icons there were scores of candles that had been lit by other worshipers. As the flames flickered their light on the statue of the suffering Christ, I also noticed the candles swelling in blisters. In a flash I remembered Benedict, a Catholic, and how he must have come upon the image of the bubbling candle and the body in pain. As Murugesu's father said, "How beautiful! What pain!"

Conclusion

I have argued then that beauty and pain are qualisigns and sinsigns respectively. In Peirce's scheme, a qualisign is a semiotic "first." It is a felt quality of experience in toto – present, immediate, uncategorized, and prereflective. Qualisigns do enter into cultural processes of significance when they are embodied and interpreted and where judgments and evaluations are made. Pain too has a present, immediate, uncategorized, and prereflective aspect to it. But this "firstness" of pain is overwhelmed by what Peirce calls "secondness" – the experience of radical otherness in which ego and non-ego are precipitated out against each other in unique and absolute opposition, where the oneness of firstness is reduced to the *hic et nunc* of actuality. "Thirdness" is Peirce's phenomenological category of mediation, mainly through language and culture. It is in thirdness that semiosis resumes its continuity, its movement. Thirdness is the domain of meaning. However, the qualisignificant and sinsignificant origins of beauty and pain respectively are such that even when they do become available to language, the traces of their sojourn in firstness and secondness remain fresh and vivid. Their emergence into general semiosis differ. In general, beauty unfolds generously and with ease. Even at the brink of language, pain lies stuck, individuated and arrested as a sinsign. Pain can be freed, however. In Sri Lanka, a culturally recognized means of effecting such a freedom entails the experience of remembered terror. Terror remembered, disarticulates and de-individuates. Terror shatters pain, and in so doing makes it available for union with beauty. It is as if pain returns to being a qualisign and then, aligning itself with beauty, rides on metaphors and icons of affect into freedom, the enabling semiosis of language and culture.

NOTES

1 A physician belonging to the indigenous, Siddha school of medicine.
2 This chapter is based on fieldwork carried out in Sri Lanka, India, Britain, France, and the Netherlands at various intervals since July 1983 through December 1990.
3 In keeping with conventions of citing from *The Collected Papers of Charles Sanders Peirce*, edited by C. Hartshorne and P. Weiss (Vols. I–IV, Belknap Press, 1931) and A. Burks (Vols. VII–VIII, Harvard, 1958), the number to the left of the decimal point indicates the volume and that to the right indicates the paragraph.
4 One was a Tamil militant, the second was a former member of the Special Task Force, and the third was a retired officer of the Sri Lankan army.
5 Peirce (1931: 2.245) defines a sinsign as "an actual existent where the syllable *sin* is taken as meaning 'being only once', as in *single, simple* Latin *semel, etc.,*" in short, a sign embodied singularly and uniquely.
6 Briefly, an iconic sign is a sign in which the representamen shares in the quality of the object.
7 A brief chapter such as this is not the appropriate place for me to develop a theory of aesthetics based on the Peircean notion of iconicity. Suffice it to say that when "beauty" itself is referred to as an "object," what needs to be remembered is that it is an "object" in the semiotic sense; i.e., not a thing in itself or an end point but something that has its own constitutive semiotic genealogy which, like all semiotic objects, is Janus-faced, both representing and being represented. Among other things that distinguish an iconic sign – say, from an indexical or

symbolic sign – are that (a) while all signs are both self-representing and other-representing, iconic signs privilege the former over the latter; (b) iconic signs are not only multivocal but are, contra symbols and indexes, unmotivated towards univocity; (c) iconic signs are paradoxical in that while their constituents may be "composed" into a whole, the harmony thus achieved is a precarious one. In addition, (b) and (c) together go into creating an aesthetic object that may be best described as throbbing with uncertain potentialities. Also see, for example, Alain Rey (1986).

  8 One reader wondered whether this initial reticence was evidence of their defense against reexperiencing pain. In the majority of the fifty or so torture victims I interviewed and reinterviewed several years later in settings of greater acquaintance and trust, this did not seem to be the case.

  9 On the use of language that conforms to Ezra Pound's dictum: "the sound must seem an echo to the sense," see Bruce Mannheim's essay by that title (1991); also see Paul Friedrich (1979) and Margaret Egnor (1986).

10 EPRLF, TELO, PLOT and LTTE are either initials or acronyms of the various Sri Lankan Tamil separatist militant groups.

REFERENCES

Alberti, Leone Batisti (1966) [1485] *Ten Books on Architecture*, James Leoni, trans. New York: Transatlantic Arts.

Chettiar, A. C., ed. (1963) *University of Madras English – Tamil Dictionary*. Madras: University of Madras.

Cohen, Ted (1982) Why Beauty is a Symbol of Morality. In Ted Cohen and Paul Guyer, eds., *Essays in Kant's Aesthetics*. Chicago: The University of Chicago Press.

Egnor, Margaret (1986) Internal Iconicity in Parayar Crying Songs. In Stuart Blackburn and A. K. Ramanujan, eds., *Another Harmony*. Berkeley: University of California Press.

Feld, Steven (1982) *Sound and Sentiment: Birds Weeping, Poetics, and Song in Kaluli.* Berkeley: University of California Press.

Friedrich, P. (1979) The Symbol and its Relative Non-Arbitrariness. In *Language, Context and the Imagination*. Palo Alto: Stanford University Press.

Kant, Immanual (1951) [1799] *Critique of Judgment*, J. H. Bernard, trans. New York: Haffner Publishing Company.

Mannheim, Bruce (1991) The Sound must Seem an Echo to the Sense: Some Cultural Determinants of Language in Southern Peruvian Quechua. In *The Language of the Inka Since the European Invasion*. Austin: University of Texas Press.

Munn, Nancy (1986) *The Fame of Gawa: A Symbolic Transformation in a Massim (Papua New Guinea) Society.* Cambridge: Cambridge University Press.

Peirce, Charles S. (1931) *Collected papers*, Vol. II, Bk. 2. Cambridge, MA: Harvard University Press.

Rey, Alain (1986) Mimesis, poetique et iconisme. Pour une relecture d'Aristote. In Paul Bouissac, Michael Herzfeld, and Roland Posner, eds., *Iconicity: Essays on the Nature of Culture*. Tubingen: Stauffenberg-Verlag.

Scarry, Elaine (1985) *The Body in Pain*. Oxford: Oxford University Press.

Shue, Henry (1978) Torture. *Philosophy and Public Affairs* 7 (Winter): 124–43.

Wagner, Roy (1985) *Symbols that Stand for Themselves*. Chicago: University of Chicago Press.

11   Rape trauma: contexts of meaning

Cathy Winkler (with Kate Wininger)

Rape trauma is disruptive of one's life, and is a disruption that might appear as meaningless emotions surfacing without reason. Rapists are the initiators of that trauma by forcefully penetrating victims physically, sexually, emotionally, and mentally. Their methods of torture implant landmines of horror in the bodies of their victims. After the attack(s), victims feel these landmines in terms of trauma – frequently a trauma unnoticeable by most people. This trauma surfaces as a feeling of separation between the mind and the body; this separation is trauma felt viscerally and unexplained mentally. In addition, rape-like comments by friends, family members, and institutional representatives instigate another type of trauma that may also set off those unseen and buried mines. I argue that an activist-survivor-victim's ability to interpret the meanings of the contexts[1] behind these traumas begins to diffuse the rape trauma and bridges the gap previously felt between the mind and the body.

Rape trauma syndrome

Burgess and Holmstrom (1974) were the first to define the concept "rape trauma syndrome." They explained the emotional chaos and symptoms from the rape attack that keep resurfacing in the victim. These symptoms demonstrate the evidence of rape trauma. In the last twenty years, researchers have added to this list of trauma symptoms that meet the now approved criteria for the diagnosis of Post Traumatic Stress Disorder (PTSD) (American Psychiatric Association 1987):

worry, exhaustion, shame, anger, withdrawal, helplessness, vaginal irritation, guilt, suicidal traits, revenge, fatigue, insomnia, decreased libido, lack of sexual desire, feelings of worthlessness, self-blame, depression, stuttering, introversion, sleeping and eating disorders, intense exhaustion, addictive traits (e.g. alcohol and other drugs), psychosomatic disorders (e.g. startled awakenings), flashbacks, moodiness, hysterical crying and screaming, phobias, nightmares, confusion, restlessness, tenseness, irritability, fear, overreacting, shakiness, paranoia, jumpiness, feelings of unreality, inability to concentrate, paralyzing anxiety, nausea, psychosexual function, and anxiety reactions.[2]

According to Burgess and Holmstrom, rape trauma syndrome develops as follows: "This syndrome has two stages: the immediate or acute phase, in which the *victim*'s life style is completely disrupted by the rape crisis, and the long-term process, in which the *victim* must recognize this disruptive life style" (1974: 3), and it "is the main *feeling* of fear that explains why *victims* develop the range of symptoms we call the rape trauma syndrome" (1974: 39, emphases added).

The authors of *The New Our Bodies, Ourselves* offer a comparable formulation of a therapeutic resolution of rape trauma: initial responses, the full impact of the physical rape pain that occurs in a safe environment, a temporary period of calm readjustment, and discussion of the raped person's deeper feelings (Boston Women's Health Book Collective 1984: 103–4).[3] Koss and Harvey (1991), however, describe recovery explicitly as a *victim-to-survivor* process with six components:

1 Memory: The individual has *control* over the remembering.
2 Integration: Memory and affect are *joined*.
3 Affect Tolerance: *affect* is no longer overwhelming.
4 Symptom Mastery: Symptoms recede, are more tolerable, and *predictable*.
5 Attachment: The individual is *reconnected* with others.
6 Meaning: The individual has *assigned* some tolerable meaning to the trauma and to the self as trauma survivor. (1991: 176–7, emphases added)

From victims' accounts, we learn in detail how the rapist's invasion splits the victim's self from the bodily experience, and how the victim is part of the rape attack.

I felt like I was in a corner of the room watching the bed from a distance. (Warshaw, in Warshaw and Koss 1988: 5)

I left my body at that point. I was over next to the bed, watching this happen . . . I wasn't in the bed . . . I was standing next to me and there was just this shell on the bed . . . When I repicture the room, I don't picture it from the bed. I picture it from the side of the bed. (Maggie, in Warshaw and Koss 1988: 56)

During the attack, while a rapist controls our bodies, but not our minds, we are already survivor-victims:[4] we saved our lives, we saved our sanity. Confronted by rape myths[5] or institutional assault,[6] survivor-victims can alter the context of victimization. Decision-making rights and freedom to make choices transform survivor-victims into activists. Activist-survivor-victims, then, are those people who claim, create, and define the meanings of the contexts – whether of suffering or not. Survivor-victims become activists when they resolve the interplay of meanings between the mind and the body.

Embodiment of knowledge

The paradigm of embodiment stresses that "the body is ... the *subject* of culture, or in other words is the existential ground of culture" (Csordas 1990: 5). This perspective allows us to look at the body as "socially informed," in the present case by the "structuring action" (Bourdieu 1977: 124). In other words, *trauma* is a subjectively structured form of knowledge.

This paradigm "asks how cultural objectifications and objectifications of the self are arrived at in the first place" (Csordas 1990: 34). The body becomes "a field of perception and practice" (ibid.: 35). Rapists who traumatize the victim's body objectify the body, and this objectification results in the victim feeling as if there is a separation between the body and the mind. Trauma then surfaces – sometimes without meaning, sometimes without warning, and sometimes as a shattering feeling. The body feels like an object, and may feel in some ways separate from the mind. The feeling is one in which the mind and body act together, but in a manner that feels disjunctive – unlike a healthy situation in which the body and the mind act in unison.[7]

The paradigm of embodiment underlines the point that reality is experienced bodily: the rapist's physical force binds his victims, his[8] sweat falls onto the victims' faces, his odors pervade the area of the attack, his saliva saturates the body parts of his victims from the face to the genital area, and his penis slams repeatedly against the victims' bodies. The meaning of trauma must be interpreted not only cognitively and affectively but sensorially. Trauma dissociates the body's feelings and senses from the interpretation, and, for traumatized people, there are contexts in which the visceral reactions are incongruent with the interpretation. In this chapter, I examine these overlapping contexts and suggest how they are loci of meaning for reintegrating self and bodily experiences.

Visceral inquiries

The questions for my research for this chapter began when someone kicked me awake and I heard the words:

"Wake up. Get up."
"Oh God! That man – standing over my bed, standing over me – wants to rape me. This is rape. He hasn't raped me yet, but already I feel that hell."

Thereafter, more unexplained emotional reactions occurred.

During the attack, the room filled with the odors of urine and excrement, odors that I thought came from my dog. Yet, no evidence of those existed, as noted by the police later.

An hour and a half after the attack, I was at the Rape Crisis Center for medical treatment, wondering what made the rape that I had just lived through feel like a hell on earth.

The following months resulted in a Star Trek Spocklike state for me – emotionless yet logical. While I was on street surveillance, the police detective called a man over to the car. His features did not match my memory of the rapist's face. As his body came physically closer to the car, the same rapelike fear that had enveloped me the night of the attack surfaced: feelings of wanting to run away from that unexplained terror seized me but feelings of paralysis bound me immobile to the car seat. I didn't know what I wanted to run away from because the horror feelings were inside of me.

Six months after the attack, still emotionless because of the rape, I began to calmly read an account of a murder-rape and suddenly the words of that rapist caused screams to cry out from my mouth and my body to evacuate its contents. (Winkler, n.d.)

When I returned to my home with the police, the lack of evidence of the odors that I had described in the police report immediately revealed to me the reasons for that experience: the odors matched my disgust for the rapist and his acts. But what were these other happenings? Why did these other visceral reactions occur? What was I experiencing? How could I understand these incongruities between the feelings in my body and the thoughts in my mind?

Rape victims are silenced both by the pervasive and invasive blame-the-victim perspectives and by pity responses that include denial of rape vulnerability by those inexperienced in rape. My experience as an activist-survivor-victim opened up opportunities to lift this silence in three ways: (1) to generate an insider's view analytically as an investigator-victim; (2) to bridge the gap for informants to become investigator-informant-victims; and (3) to revamp my approach to allow an informant to transform her/himself[9] into an activist.

1. Unlike most anthropologists whose decision to study a research area is by choice, both in context and in subject matter, for a crime victim, there is no choice. Yet, it is in that lack of choice that a crime victim fully comprehends the status of that subject matter. Nevertheless, an investigator-victim does not lose her/his ability to be an observer because of the criminal context. Rather, a life-threatening situation can heighten one's power of observation and accentuate one's ability to decipher the meanings behind a criminal's words, and in this case, heinous acts.

The investigator-victim has that distinctive blend of subjective–objective perspective (Du Bois 1967). As one subjected to rape, my objectivity did not become ruled over or ruled out by the rapist; the rapist did not completely make me into a subject of *his* rape. Like other victims, I knew that same unexplainable terror (Sheffield 1987), but unlike many victims, I had a background in cross-cultural experiences[10] that enabled me to analyze the meanings, interactive processes, and trauma reactions. More importantly, the experience of rape aided me as an investigator to frame questions more appropriately with investigator-informant-victims, to develop an analytical approach for this type of data, and to search out questions previously unsolicited by nonvictim-investigators.

2. Investigator-informant-victims are more than describers of data; they are the interpreters of the rapist's acts and meanings. While the investigator of the research can provide them with a framework for analysis, the distinctiveness of the rapes and the rapists makes it necessary for survivor-victims themselves to become researchers into the meanings of the rape trauma experiences. Investigator-informant-victims must be the interpreters and analyzers of the rape trauma because, first, the initial context of the trauma was a secluded, inaccessible area known only to them; second, the inflictors of the torturous emotions are rapists whose identities remain, by and large, anonymous and thus unknown to investigators; and third, the silencing of the victim – part of the rape trauma – must be broken by re-establishing investigator-victims as the voices of their identity and with the meanings they associate with that attack. The meanings are buried in the mind–body disconnectedness. These data can not be seen, observed, or heard by outsiders. These are experiential data, knowledgeable only to survivor-victims.

→ 3. Structured interviews are eschewed.[11] Since rapists have already tried to take away our decision-making rights and choices, interviewers must not duplicate this type of behavior. In my study, investigator-informant-victims select the time, place, and duration of our exchanges, and they decide on the extent of my participation. In this research, most exchanges have lasted three to four hours. If a predicament has arisen to alter the time of the interview, the investigator-informant-victim has chosen the next time that would accommodate both of us. The sites of exchanges of information chosen have varied from my office to restaurants and parks, and have also included telephone conversations and letter writing. Finally, investigator-informant-victims have had the right to edit the work and delete portions of their interview. Word choice and phrasing have been an interactive process between us. This editing has not only clarified points important to the investigator-informant-victims, but has given them decision-making rights. The ethnography thus holds the ingredients of the thoughts, feelings, and experiences as only known by them.

For other investigator-victims and myself, trauma had blocked our understanding and muted the severity of those unexpected blasts of emotions. Continuing to live in minimally supportive contexts, our minds pursued those protective strategies, but our bodies became the interpreters of contexts that were dangerous or socially confrontative by warning us through shakiness, nausea, and waves of fear and trembling. The analysis that follows is based on in-depth research with fourteen people. We identified five contexts[12] that allowed us to organize the meanings of our bodily experiences. Each different context frames a set of meanings and becomes an additional source of trauma, compounding the occasions of visceral and cognitive interpretation.

Attack-like context

Virtually any element of the attack context can trigger the feelings felt by investigator-victims during the attack. For instance, a rapist seized Shirley[13] from behind and held a knife to her neck as he dragged her into the alley. For years, any touch to that part of the neck – from a gentle caress to a sweet kiss – still released those feelings felt during the attack. In my case, the rapist attacked at night in my bedroom. To avoid those nocturnal feelings associated with the attack, a night-light was my comfort. But one time, in the middle of the night, and with a sudden adrenaline rush, I woke up and jumped out of bed. The light had burned out. The lack of light at night – a point noted by the rapist – had placed me back in the context of the attack. The adrenaline-high, the agitated state, and the sensation of bewilderment were present. Yet, in this case, the feelings subsided quickly once the reason for the surfacing of those horror reactions became clear. Meaning defused the landmines of emotion set off by the burned-out light. Also, a spare bulb helped to reinstate slumber.

Meaning in attack-like contexts

Similarity between the rape attack context and the nonrape attack context is the reason for the surfacing of those trauma feelings. While this perspective seems reasonable, one wonders why a context that is largely safe and secure still generates rape trauma? Do rape survivor-victims want to hold on to those memories of horror? We must place ourselves in the position of survivor-victims.

With the rape trauma buried inside me along with the fear of the rapist still alive and free to attack again, my body awakened me in order to protect me. Time and space were variables still necessary for healing, and these trauma responses were a means of alerting me to distance myself from potentially dangerous contexts in order to seek out contexts of greater safety – away from the rapist and his tactics.

Rape-myth or second-assault contexts

Rape myths are statements – intentional or not – against survivor-victims. These statements, which are discordant with the experience of survivor-victims, are taken for granted by those who repeat them. Rape-myth statements are context-specific. The context of the rape attack is usually the source of rape myths. If the survivor-victim met the rapist in the bar, then s/he is blamed for going to that bar, for talking to a person in that bar, or for getting a ride home from a seemingly kind person. Uneducated about rape, some people make rude statements fabricated from the context of the attack.

Rape myths are part of what Williams and Holmes (1981) call the "second assault." The second assault, or "social rape" as I have called it (C. Winkler 1989), is the accumulation of negative responses (Williams and Holmes 1981). These negative responses, while sometimes indirect and unbeknownst to the speaker, repeat the verbal attacks of the rapist. Discriminators, whether expressing themselves in the form of rape myths, racism, sexism, or ageism, alter their context of reference over time. This alteration then disguises their comments against survivor-victims. I call this process the *chameleon of disguise* (C. Winkler 1989). Persons raped[14] and in a state of silence are told "You should *not* have gone to the bar," and yet in my experience as an activist-survivor-victim, I was told "You should not speak up, and you should not tell people about *that*."

If the survivor-victim is able to thwart and prevent the characteristic rape myths by strategically arguing against them as I did, then discriminators find other contexts in which to thwart and blame survivor-victims. Initially, people tried to blame me for living in that neighborhood or for not having bars on my windows, but my familiarity with these types of attack-associated myths helped me quickly to dispel them. As a result, people altered their rape myths to fit in with another context; in my case, the discriminators argued against my right and ability to speak about rape. My openness became their target.

For example, I explained my week's absence to my students with a twenty-minute talk about the attack and the bruises on my face, answering their questions in a straightforward and honest manner. Other than that, the course centered on the original content. But the discriminators had ideas of their own. Prior to my knowledge of their comments, I began to experience a shakiness in my legs as I approached the departmental office. Some of my colleagues had, in fact, begun to argue that I was teaching courses on rape. The actual content of my courses became irrelevant to them. Significantly, all the women faculty and staff except one workstudy student had been previously raped, and the bruises on my face reminded them of their own past pain.

In another instance, my colleague Jerome (see note 13) was organizing a bulletin board on the faculty's research and publications. One month after the attack, he confronted me for being one day late in submission of materials. Discussion with Jerome of my priorities – stopping the rapist, living in hiding, relocating, problems with bill payments and bank statements, and hiring a private investigator – was fruitless. Jerome was furious: he screamed at me for my failure to do my work. During his harsh reproaches, he suddenly became quiet. Since the attack, I had not lost my temper, I had never become upset around the office, and I had not reacted – in a crucially traumatic situation – unreasonably. Of course, I was still in a state of numbness. Yet he, who was untouched directly by the crime, could not control himself and could lose his temper over a bulletin board. I found out later that the reason for his reaction was his difficulty in accepting the trauma of a stranger-rape that his wife had also suffered years before he met her. Knowledge of my agonizing torture meant he had to confront the pain his wife had once endured.

Common statements in our culture sometimes parallel the words of the rapist, and thus bring the context of the attack to the surface. The rapist announced my wants and dislikes during the rape – "You like this" and "You don't like this" – just as others decided and then announced my reactions – "You're upset about this" and "You're angry about that." Neither would let me answer.

Just as the rapist felt that he had a right to determine my thoughts and feelings, these people felt they had that same right. An administrator of the university refused to change the keys to the office door despite the fact that the rapist had stolen my keys and knew where I worked. She announced many times: "You are paranoid." In another instance, when I had told one of my male colleagues that I had slept poorly because of a new mattress, he announced that the "real truth" was that I was suffering from nightmares because of the rape, but was refusing to tell him the truth. My meanings were discarded. While some rapelike statements can be handled by survivor-victims, the constant bombardment of these statements from people becomes a second assault or, more precisely, a second rape.

Meaning behind rape-myth or second-assault context

One rapelike hateful comment might be forgotten by survivor-victims, but when our *community* of supposed supporters, the people in whom we believe, are the ones who formulate negative statements that they throw at us from every corner of "support," then we suffer a second assault. These statements are a relentless verbal attack. People who make these statements deny how we feel and blame us for the rape. The objective approach of "supporters" who provide answers for us separates ourselves from our self-objectification, and this separation is yet another set of meanings traumatized into our bodies.

Like the physical rape attack, then, the second assault, or social rape, threatens our identity and our definition of self. We suffer another rape. The meaning behind the second-assault comments becomes the horror that our bodies react to. In a state of trauma, emotional tremors surface in our bodies to protect us from these people.

Guilt context

Guilt is one of the greatest silencers of survivor-victims. It covers up the crime, it protects the criminal, and it places survivor-victims in a cocoon of fear. The wall of guilt inhibits expression of oneself and inhibits investigators from deciphering the contexts that generate rape trauma. Guilt is a tool of psychological torture against survivor-victims initiated by the rapist – "You wanted it" – and provoked by supposed supporters – "Why did you let him do it?" This is one area of rape trauma difficult to comprehend.

Donna Queensley[15] lived behind that wall for years, and the rapist David built that wall. He began his amorous and chivalrous approaches to Donna when she was 17 years old and just learning the meaning of sexual expression. His sexual effect alternated between pleasure and pain. During those good times, he affectionately and amorously made love with Donna; at other times, he raped her. His addiction for these love/hate sexual encounters became her blight. After David, Rob – on the first date – raped her. Yet, Rob was easy to walk away from because there never was any tenderness to mistake in his actions of force. Donna thus has experienced multiple types of rape: first, she was in the love/rape relationship with David; second, she suffered the terrorizing and physically threatening rape by Rob. Despite therapists, counselors, and mental-hospital professionals, Donna experienced years of mind–body confusion. But her inner sensibility and strength challenged the trauma of the rapist-built walls of guilt that had silenced her.

During those times of self-blame, Donna lived with incongruous sensations between her mind and body. In the act of sexual intercourse with other men, Donna felt her body react physically, but without those feelings of love. In the presence of her parents, Donna's thought processes became interrupted with a loss of concentration and lack of desire. Waves of dread swept over her.

David the rapist had attacked Donna's existence and separated pleasure from Donna's sexual self. This split between emotional feelings of love and the physical sensations of sexual love is mendable. But his destruction and attacks on her also triggered episodes of manic depression. While David did not cause the *condition* of manic depression, Donna notes that David-the-rapist was the *initiator* of her episodes. In other words, the trauma from David's rapes aggravated her psychiatric disorder.

Donna felt the trauma around her parents; it consisted of nervousness, shortness of breath, waves of dread, heaviness in her legs, and a pain in her stomach. This guilt trauma centered on issues of premarital sexual intercourse, her failure to fulfill the role of a model daughter, and feelings that she had committed an awful wrong but without understanding the wrong, nor that the wrong was David's rapes. Moreover, Donna felt that she was not acting like an adult and not taking responsibility for herself as an adult. She felt that when she returned home after an evening in which David had attacked her, her parents would discover that she had had "sexual relations" with him.

Using my analysis of context and meaning to interpret guilt trauma, Donna spent two weeks in a state of emotional chaos deciphering the meanings. She became the investigator. Although the examination of that horror almost broke up her relationship with her fiancé Victor, a wonderfully sweet and gentle person, Donna had a determination to extinguish that trauma from her life. Her ability to comprehend the trauma ended the guilt, and her analysis of the meanings of the context of the traumas stopped those feelings of self-blame during sexual intercourse and when she was with her parents. Thereafter, sexual intercourse became a pleasure for her. The guilt in the presence of her parents likewise ended. Donna didn't have to protect herself anymore because *her* meanings returned her identity – an identity with a foundation greater than the rapist's terror.

Meaning in guilt context

Donna's first signs of meaning were generated by her body through feelings of fear and lack of mental concentration. The rapist had instilled the guilt by his rapes, and the presence of Donna's parents reaffirmed another form of that guilt. The purpose of the guilt, the assault on her self-concept, was to keep her quiet; yet her body, separated from her self, screamed for affirmation. Her connection between her silence and the trauma-shouts of her body in those contexts revealed to her the meaning of David's acts as rape and the meanings of her parents' presence as authority figures. Her deciphering of these meanings allowed Donna to feel, for the first time, a unified connection between love and sexual intercourse.

For a decade, Donna had experienced guilt trauma when she had sexual intercourse or when she was in the presence of her parents. The bifurcation between her mind and body had continued despite intensive therapy. Upon completion of our exchanges, Donna's guilt traumas ended. She felt that the termination of her traumas was a result of this research approach and methodology. When she read her investigator-informant-victim perspective in this chapter, she wrote the following words:

I have carefully read the paper. The first thing I said to myself when I finished was "Wow!" You totally captured my thoughts, feelings, and experiences. As I read it, I could relive the feelings I had at the time; so it must be a good article. I can tell that

you really put your heart into it. I appreciate its accuracy. I consider our talks at ——
to have been a very healing and liberating experience. Now when I make love, I feel
no guilt and I have total trust. (Personal communication)

Rapist-identified context

During the months following the rape attack, the police detectives took me
around the neighborhood on surveillance, sometimes calling men over to the
car so I could observe them closely. In observation of all men but the
following case, my emotions were numb. With the closeness of one man,
whom I could not visually connect as the rapist, explosions of terror similar
to those that I experienced on the night of the attack permeated and swelled
inside my body (Weis and Borges 1973).

This is what happened. The man was standing about ten yards from the
police car, nonchalantly talking with a friend. The detective called him over.
I superimposed my mental image of the rapist over the physical image
walking in my direction. Once within my personal boundaries – a range of
six to nine feet – the rapist's presence detonated the terror. This emotional
force was unexplainable. I wanted to flee the scene and to run away from the
demon. Where would I go? The demon was inside me.

Recognition of that man's face as the rapist's was not possible. Similarity
of facial features did not equal an absolute identification, and I said to the
police detective: "That's not the man." Of interest are the next sentences
that came forth from me as I viewed his back.

But, remember that body. That is the same body as the rapist and the shape of his
head is *identical*.

My body's sensibility counteracted part of the terror: I could at least match
the body form of the man on the street with the body form of the rapist. A
further intriguing point is that my mind matched the front shape of the
rapist's head – the only side of him that I saw during the attack – to the back
image of the man as he walked away. There was no doubt in my mind as to
the similarity of those body forms. This is the one and only instance in
which I made a statement of similarity to the police – and with confidence.
At that point, no explanation for my failure to identify the face of the rapist
or for my success in identification of the rapist's body was available.

Later I asked myself: Why hadn't anyone told me about bodily identification? Why couldn't my mind break down the protective barrier of terror
to enable me to recognize that man as the rapist? Were the rapist's threats
and methods of traumatizing me so effective that my mental intent of
protecting myself could not let in the shattering truth of the horror that
occurred when the rapist bludgeoned and rammed my body over and over
and over?

The explanation for my bodily and mental responses was best clarified by
Kate Wininger. Worrying about the experience of identifying the rapist, I
asked her what happened during the police line-up in which she identified the
rapist who had attacked her: *How* did you identify the rapist? Kate explained:

I had gotten a good look at the rapist in spite of the fact he had put clothing over my
face. The presence of the gun, even though I didn't believe it was real, and being
strangled until I became unconscious along with the rapist's threats to come back and
kill me and my husband, all made the trauma more real.
  Nevertheless, I still felt competent I could identify him. In the police line-up, the
men singly walked into view. I was not afraid, and had no reason to feel fear because
these men were behind a one-way mirror. Yet, one of them had an impact upon me.
When he entered my vision, my body first identified him in the line-up. It was *a purely
visceral response, a return of the fear and revulsion*. My body told me immediately that
he was the rapist. After my body's split-second identification, I then mentally recalled
that man's face as that of the rapist. (Emphasis added)

In analyzing Kate's context of identification, she was in a position of safety.
As she described it, the rapist was behind a one-way mirror. He didn't know
who was identifying him because there were a number of other survivor-
victims to make the identification. In this context of security, Kate could let
down her protective barriers.

In my case, pieces of the puzzle began to fit together months later. The
artist's composite of the rapist was identical to the face of the man on the
street, and this irrefutable fact had kept the police after him. But, without my
ability to identify him, the police had no evidence that was legally acceptable.
The discovery of DNA Fingerprinting (Lewis 1988) convinced everyone
that the unidentifiable man-on-the-street and the rapist were the same
person and revealed to me the meaning of my reactions during street surveillance. Through the re-enactment of the rape terror, my body informed me
that the man-on-the-street was the rapist (compare Csordas 1993 on
"somatic modes of attention"). My mind continued to protect me because
the threat of another rape was a major issue: the rapist could still run and hide
from the police.

Meaning behind rapist-identified trauma

In rapist-identified contexts, the trauma engulfs the victim in the same
paralyzing manner as during the rape attack itself. The identification of
trauma with the attack – in its intensity and pervasiveness – is a means by
which our bodies give us information. Our bodies provide interpretations
by warning us of the pending peril because of the presence of the rapist or
another similar abuser. This warning is a notice to remove ourselves from
that dangerous context.

The paralyzing fear that has entrapped our minds is superseded by our bodies' screams of trauma. The rapist's objectification of my body pounded in a set of meanings that announced these separate realities of the body and the mind through trauma. The all-encompassing trauma reactions are our bodies' way of announcing in a clear and articulate manner that there is a meaning to this context undecipherable at this time by the mind. Counselors and therapists had explained to me that these body traumas were like emotional upheavals that I would get over. To them, tremors were experiences to live through and forget. These traumas are *not* meaningless residual pain from the attack. Rather, these traumas contain, in this instance, alarm meanings that activist-survivor-victims need to learn about and interpret.

Blame-denial contexts

At the Rape Crisis Center, I met the Rape Survivor Advocate. She immediately pulled me from the cliff of insanity and treated me like a person who could deal with that bludgeoning pain. I told her that the rapist had wanted to kill me. She assured me that all rapists threaten their victims, but that few carry through with their threats; besides, the rapist had no weapon. I accepted her position that the rapist's threats to my life were my fabrication, and assumed that my shock must have distorted the reality of the rape attack. Nevertheless, the rape attack was like a dark cloud of horror over me – despite my state of numbness – and it was clear that the cloud had to be dispersed. After three months of numbness and another three months of short spells of crying in private, my body provided the answer. On 26 March 1988, the *New York Times* ran an extensive article on the well-publicized prep-student rapist-murderer. I read this – as I had read every other article on rape in the previous six months – to learn about and confront the trauma. But this time my reaction was different.

This rapist had offered the woman a ride home and then taken her to Central Park to rape her. The points of the article were not of particular significance to me, but the words of the rapist had an astounding impact upon me:

Mr. Chambers, who is 21, told the authorities after arrest – in a written statement and a videotape that was played repeatedly at the trial – that Miss Levin died accidentally during sex. He said she had pursued him at an Upper East Side bar on the night of her death and had finally talked him into going to Central Park sometime after leaving the bar at 4:30 am. Once there, he said, she went "insane," scratching and biting him and finally squeezing his testicles so painfully that *he grabbed her "instinctively" from behind the neck to stop the pain*. He stressed on tape that *he never meant her harm*. (Johnson 1988: 32, emphases added)

The meaning of the rapist Chambers' words matched the meaning of the rapist's words spoken to me during the attack: "It's your fault that I hit you. I didn't want to do it, but you made me beat you up. *I had to stop you*" (C. Winkler n.d., emphases added). Both rapists blamed their victims. Ms. Levin did not cease protecting herself from the rapist Chambers' battering until she could fight no more, or until he had killed her. In my instance, if I, unlike Ms. Levin, had not quit fighting and had not completely and absolutely given in, the rapist Redding would have also ended my life. Rape was my escape from physical death.[16]

As I read the rapist Chambers' words, I spontaneously screamed in a death-defying manner and my body evacuated its contents. In an explosive emotional experience, those murderous moments became evident. My body told me that I was now safe to confront that meaning because I was alone, without people to deny my experience. The meaning that I had to accept was that I had had a confrontation with physical death. Almost two more years passed before I read the rest of that article.

Meaning behind blame-denial contexts
While many people blame survivor-victims for the rape attack, and then also tell them that they are in denial, I argue that the people around me were in denial and that I had initially believed their words of denial. The bomb of emotions was my body's means of breaking the barricade of ignorance imposed by my "supporters." The Rape Survivor Advocate and others objectified my meaning and that objectification inserted yet another trauma inside me. Somehow my body knew that this safe and secluded three-day weekend at home was a place to accept that reality.

An emotional landmine had exploded and revealed the meaning of rape. For me, rape was an encounter with physical death. It was also the rapist's attempt to socially murder me by eliminating my ability to act, decide, or define myself other than as the rapist dictated (C. Winkler 1991).

Unscrambling trauma

Rape trauma is experientially unique. In my case, the lengthy numb period allowed me to differentiate the contexts of the second assaulters from the context of the rapist. DNA Fingerprinting, which provided positive identification of the defendant as the rapist, legitimized my bodily reactions of terror during street surveillance as not false or manufactured but a viable trauma inaugurated by the rapist. The absence of feelings of guilt, denial, blame, and vulnerability – common for some survivor-victims but not present in my case because of self-education during the attack, and self-defense against the rape myths – was not part of my trauma, and did not become intermixed with other types of trauma. Nevertheless, there were times when my interpretation of the disruptive feelings – whether provoked by the rapist, the second-assaulters, or others – was barely decipherable. One difficult interpretive period contained a mixture of assaults from social rapists, sex discrimination, and loss of my job as a result of the attack, and the realization that I knew the face of the rapist while my mind would not let me identify him. These periods of multiple overlapping traumas scrambled the meanings.

Historically, Burgess and Holmstrom's (1974) research is exemplary in that prior to their studies the concept of rape trauma did not exist. Now, almost two decades later, rape-trauma experts argue for an understanding of trauma as a process delineated by phases and symptoms. This typological approach, I argue, is an insufficient therapeutic approach for activist-survivor-victims. For many survivor-victims the intertwining of trauma contexts impedes her/his ability to translate the body's language. Even in a case such as mine – in which each of the traumas contained its own distinctiveness, not mixed with other trauma ingredients – discovery of the meaning of that trauma wall was a challenge.

Understanding context and meaning helps to refine our understanding of rape trauma. It is also an aid to altering the rapist-defined position of victims in previous research studies *into* the position of activists, where they define the meanings of the rape attack and the trauma. This approach, then, helps to diffuse the impact of rape trauma and, as such, is an approach that is more than explanatory; this approach is one of activism that gives survivor-victims the right to define their own identities.

In attack-like contexts, the severity of the rape still binds survivor-victims who need time to heal, and who, because of social and legal rapes, are not in a safe context. The body warns of other contexts that do not have the prerequisite security "locks." With rape myths, the disguised blame-the-victim statements are another social or second rape, and the body forewarns survivor-victims of unfriendly and nonsupportive comments. Guilt trauma, which can wall survivor-victims from the world of people, reinforces the authority of the rapist and his control, and an understanding of its sources is a means of jolting survivor-victims into reassertion of their rights. Rapist-identified trauma, a type not recognized as legal evidence, points out the rapist or other abusers to survivor-victims. Lastly, blame and denial trauma, which isolates survivor-victims, formulates an incongruity that can ultimately validate survivor-victims' interpretations.

Survivor-victims' analyses of their traumas through a study of the contexts in which the traumas surface can offer meanings, and these meanings diffuse the rape trauma by bridging the felt gap between the mind and the body. Survivor-victims become activists when they interpret the meanings of rape trauma. If activist-survivor-victims interpret these contexts and sets of meanings, they can better help themselves to deal with the rocky road to recovery. I argue that the following four therapeutic functions affirm this activist approach:

1.   Protection. The rapist's physical insertion of his penis with bludgeoning force into his victims implants horror in her/his body's memory, and the memory later acts as an alarm system. As their bodies react traumatically, survivor-victims learn to guard themselves against the presence of the rapist or another abuser. The body brings to the attention of the survivor-victim, in a dramatic visceral manner, the occurrence of danger. Although the survivor-victim may not understand the meanings, the upsurges of emotions cause survivor-victims to find a safe haven, a place away from the presence of the rapist/abuser.

2.   Forewarning. Since the impact of psychological, mental, and physical terror by the rapist and other nonsupporters can be extensive, and since the impact of the terror may not be clearly perceived as a danger by survivor-victims, their bodies re-enact horror feelings to warn them of potentially unsafe contexts. Thus, this trauma is important for self-preservation.

3.   Safety. The nonrape context, while resurrecting past trauma, can also be a context of safety in which survivor-victims can deal with the trauma and its meanings without fear. The rape trauma is pronounced in force because the context does *not* contain other threats. In a safe context, s/he can let down her/his protective guard. Releasing her/his guard allows the trauma to be felt and aids in the reconnection of meanings viscerally and cognitively. It is this context of safety that enables survivor-victims to delineate and decipher contexts of nonsupport in order to reveal their translations of those threatening contexts. Such interpretive actions convert them to activists.

4.   Information. The momentous shock of the rape attack or other rape trauma may contain some devastating meanings for survivor-victims to confront. Denial of these meanings by those people around survivor-victims can further impede an understanding of our feelings. Acceptance of these meanings may be eased when our bodies shout the validity of the meanings. Visceral reactions validate and facilitate the mind's interpretations.

The analysis in this chapter defines some of the meanings of rape trauma and describes contexts experienced viscerally and mentally by activist-survivor-victims. This research, I suggest, is but one step in pursuing an anthropological understanding of violence.

ACKNOWLEDGEMENTS

Thanks to Thomas Csordas for his request of my work in this collection, his patience, his editorial insights that enhanced the writing and organization, and, most especially, his ideas on the paradigm of embodiment that provided a framework for my analysis.

I presented versions of this paper at the 88th American Anthropological Association Meetings in Washington, DC (1989) and at the American Ethnological Meetings in Atlanta, Georgia (1990). Another form of this argument is in *Many Mirrors: Body Image and Social Relations*, edited by Nicole Sault (forthcoming).

I would also like to thank Neni Panourgia for her astute comments that helped to refine the writing and argument and the reviewer whose criticisms were accurate and supportive.

NOTES

1 Peter Reynolds deserves special thanks for his review of an earlier version of this paper, but, more especially, for his advice on the use of "context" in my analysis.

2 The following are some of the researchers who have charted or discussed the symptoms of rape: Sutherland and Scherl (1970); Burgess and Holmstrom (1974); Seligman (1976); Frank, Turner, and Duffy (1979); Kilpatrick et al. (1979a, 1979b, and 1983); Veronen and Kilpatrick (1980); Kilpatrick et al. (1981); Atkinson et al. (1982); Nadelson et al. (1982); Ellis (1983); Koss (1985); Santiago et al. (1985); Torem (1986); Stewart et al. (1987); Frazier (1990); and Gidycz and Koss (1991).

3 Other researchers who present studies on the phasic interpretation of rape trauma are: Sutherland and Scherl (1970); Kilpatrick et al. (1979a); Becker and Abel (1981); Atkinson et al. (1982); Ellis (1983); and Koss (1985). Different treatment methods are found in Spring (1977); Frank et al. (1979); Kilpatrick et al. (1979a, 1979b, and 1983); Kilpatrick et al. (1981); Turner and Frank (1981); Ellis (1983); Holmes and St Lawrence (1983); Myers et al. (1984); Kilpatrick and Calhoun (1988); and Burkhart (1991).

4 In most cases, I prefer the plural of "survivor-victims" and "activist-survivor-victims" to emphasize that rape is not a one-time, singular event experienced by an individual, but that the number of people raped is sizable.

5 Some studies on rape myths are: Schwendinger and Schwendinger (1974); Feild (1978); Burt (1980); and C. Winkler (1989).

6 Literature on institutional assault includes: Holmstrom and Burgess (1978); Estrich (1987); Sanday (1990); Parrot (1991); and C. Winkler (1992).

7 Blacking (1977) argues that the mind and body act in unison and generate meaning simultaneously. In a health context – i.e., without past or present trauma, crisis, or stress – I agree.

8 For some people, the gender and sex of the rapist as male are in question. In the case of social rape discussed later in this paper, I agree. But, in physical rape, most research and victims' accounts point out only one common trait for all rapists: they are male.

9 While most people raped are women, some men likewise have suffered this horror. Male victims have asked me to recognize their plight by the careful selection of pronouns. Moreover, I argue that, while there are multiple meanings of rape for the victims, the horror, despite one's gender, is equally unjust in all cases.

10 My cross-cultural study on rape began during my research (1980–2) for my dissertation (1987) in Mexico, which centered on the changing acquisition and

expression of authority by gender in the lacquer industry. A review of the history (1940–82) on men's and women's changing gender roles demonstrated an alteration in meaning of the once-acceptable marriage practice of *robar la mujer* from the 1950s meaning "to steal the woman from her parents" to the 1970s meaning "to rape a woman." That chapter analyzed the act and meaning of rape and the community's response in favor of and supportive of the victim.

11 Many psychological studies note a high drop-out rate for their research (see Ellis 1983; Holmes and St Lawrence 1983). Such studies are based on a controlled regimen – from the time and place to the formatted, unalterable questions. For victims, such controls are comparable to rapists' controls. According to psychologists, alteration of the methodology or controls seriously affects the validity of the study, and their "objective" point of view is stressed over concern for the victims.

Within a year after the attack, I volunteered for such a study, and became a drop-out after the initial phone call. The researcher would only conduct the study in the area of Boston where *her* office was located: this was not the safest area, and it was impossible to park there. The exacerbation one might experience in just arriving at the designated place was irrelevant to the researcher, and changing to another more feasible location would, according to her, invalidate the study. While the researcher had the right to tape my responses, I was *not* allowed to have a copy of *my own words*. A copy of my words for my own therapy was against protocol.

12 This paper only explains a few of the contexts: clearly a multitude of contexts and meanings exist. Two major contexts and meanings not discussed here are vulnerability contexts and legal contexts. Briefly, the former emphasizes the lack of ability to protect oneself, and is the victim's warning system; the latter is the relived rape attack trauma in which the legal establishment's harassments (C. Winker 1992) and the methods of the legal experts are attack-like and the victim needs again to protect her/himself from them. The reclamation context is termination of trauma in which the mind and body reunite (Winkler, forthcoming).

13 This is a pseudonym.

14 The rapist's attack is a horror *survived* by the person. Yet that survival becomes further threatened when people use phrases such as the "raped person." Grammatically accurate yet victim-defining, such phrases stress the experience over the qualities of the individual. I prefer the phrase "person raped," a phrase that honors the individual first and recognizes the attack second. Perhaps our terminology contains a perspective that initially demotes the person raped, and reinforces the rapist's treatment and attitude to a victim.

15 This is a pseudonym. After hearing in class about my research and the status of the trial against the rapist, Donna volunteered to help in my work.

16 While the attack by the rapist against me contained the possibility of physical murder, I do not believe that all rapists are physical murderers (on rapists as social murderers, see C. Winkler 1991).

REFERENCES

American Psychiatric Association (1987) *Diagnostic and Statistical Manual of Mental Disorders* (3rd edition revised) Washington, DC.

Atkinson, B. M., K. S. Calhoun, P. A. Resick, and E. M. Ellis (1982) Victims of Rape. *Journal of Consulting and Clinical Psychology* 50: 96–192.

Becker, J. and G. Abel (1981) Behavioral Treatment of Victims of Sexual Assault. In S. M. Turner, K. S. Calhoun, and H. E. Adams, eds., *Handbook of Clinical Behavioral Therapy*. New York: John Wiley.

Blacking, J. (1977) *Anthropology of the Body*. New York: Academic Press.

Boston Women's Health Book Collective (1984) *The New Our Bodies, Ourselves*. New York: Simon and Schuster.

Bourdieu, P. (1977) *Outline of a Theory of Practice*. Cambridge: Cambridge University Press.

Burgess, A. W. and L. L. Holmstrom (1974) Rape Trauma Syndrome. *American Journal of Psychiatry* 131: 981–6.

Burkhart, B. R. (1991) Conceptual and Practical Analysis of Therapy for Acquaintance Rape Victims. In A. Parrot and L. Bechhofer, eds., *Acquaintance Rape*. New York: John Wiley, pp. 287–303.

Burt, M. R. (1980) Cultural Myths and Support for Rape. *Journal of Personality and Social Psychology* 38: 217–30.

Csordas, T. J. (1990) Embodiment as a Paradigm for Anthropology. *Ethos* 18: 5–47. (1993) Somatic Modes of Attention. *Cultural Anthropology* 8:135–56.

Du Bois, W. E. B. (1967) *The Philadelphia Negro*. New York: Schocken Books.

Ellis, E. M. (1983) A Review of Empirical Rape Research: Victim Reactions and Response to Treatment. *Clinical Psychology Review* 3: 473–90.

Estrich, S. (1987) *Real Rape*. Cambridge, MA: Harvard University Press.

Feild, H. S. (1978) Attitudes Toward Rape: A Comparative Analysis of Police, Rapists, Crisis Counselors, and Citizens. *Journal of Personality and Social Psychology* 36: 156–79.

Frank, E., S. M. Turner and F. Duffy (1979) Depressive Symptoms in Rape Victims. *Journal of Affective Disorders* 1: 269–77.

Frazier, P. A. (1990) Victim Attributions and Post-Rape Trauma. *Journal of Personality and Social Psychology* 59(2): 298–304.

Gidycz, C. A. and M. P. Koss (1991) The Effects of Acquaintance Rape on the Female Victim. In A. Parrot and L. Bechhofer, eds., *Acquaintance Rape*. New York: John Wiley, pp. 270–83.

Holmes, M. and J. S. St Lawrence (1983) Treatment of Rape-Induced Trauma. *Clinical Psychological Review* 3: 417–33.

Holmstrom, L. L. and A. W. Burgess (1978) *The Victim of Rape: Institutional Reactions*. New York: John Wiley.

Johnson, K. (1988) Chambers, With Jury At Impasse, Admits 1st Degree Manslaughter. *New York Times*, 26 March.

Kilpatrick, D. G. and K. S. Calhoun (1988) Early Behavioral Treatment for Rape Trauma: Efficacy or Artifact? *Behavior Therapy* 19: 421–7.

Kilpatrick, D. G. P. A. Resick, and L. J. Veronen (1981) Effects of a Rape Experience: A Longitudinal Study. *Journal of Social Issues* 37(4): 105–22.

Kilpatrick, D. G., L. J. Veronen, and P. A. Resick (1979a) The Aftermath of Rape: Recent Empirical Findings. *American Journal of Orthopsychiatry* 49: 658–69.
   (1979b) Assessment of the Aftermath of Rape: Changing Patterns of Fear. *Journal of Behavioral Assessment* 1: 133–49.
   (1983) Rape Victims: Detection, Assessment, and Treatment. *Clinical Psychologist* 36: 88–101.

Koss, M. P. (1985) The Hidden Rape Victim: Personality, Attitudinal, and Situational Characteristics. *Psychology of Women Quarterly* 9: 193–212.

Koss, M. P. and M. R. Harvey (1991) *The Rape Victim: Clinical and Community Interventions.* Newbury Park: Sage.

Lewis, R. (1988) DNA Fingerprints: Witness for the Prosecution. *Discover* (June): 42–52.

Myers, M. B., D. I. Templan, and R. Brown (1984) Coping Ability of Women Who Became Rape Victims. *Journal of Consulting and Clinical Psychology* 52: 7308.

Nadelson, C. C., M. T. Notman, H. Zackson, and J. Gornick (1982) A Follow-Up Study of Rape Victims. *American Journal of Psychology* 139: 266–70.

Parrot, A. (1991) Medical Community Response to Acquaintance Rape: Recommendations. In A. Parrot and L. Bechhofer, eds., *Acquaintance Rape.* New York: John Wiley, pp. 304–16.

Sanday, P. R. (1990) *Fraternity Gang Rape.* New York: New York University.

Santiago, J. M. F. McCall-Perez, M. Gorcey, and A. Beigel (1985) Long-Term Psychological Effects of Rape in 35 Rape Victims. *American Journal of Psychiatry* 142(11): 1338–40.

Schwendinger, J. and H. Schwendinger (1974) Rape Myths: In Legal, Theoretical, and Everyday Practice. *Crime and Social Issues* 1: 18–26.

Seligman, M. E. P. (1976) Phobias and Preparedness. *Behavioral Therapy* 2: 307–21.

Sheffield, C. J. (1987) Sexual Terrorism: The Social Control of Women. In B. B. Hess and M. M. Ferree, eds., *Analyzing Gender: A Handbook of Social Science Research.* Newbury Park: Sage, pp. 171–89.

Spring, S. (1977) Resolution of Rape Crisis: Six to Eight Month Follow-Up. *Smith College Studies in Social Work* 48: 20–4.

Stewart, B. D., E. Hughes, E. Frank, B. Anderson, K. Kendall and D. West (1987) The Aftermath of Rape. *Journal of Nervous and Mental Disease* 175(2): 90–4.

Sutherland, S. and D. J. Scherl (1970) Patterns of Responses Among Victims of Rape. *American Journal of Orthopsychiatry* 40: 503–11.

Torem, M. S. (1986) Psychological Sequelae in the Rape Victim. *Stress medicine* 2(4): 301–5.

Turner, S. and E. Frank (1981) Behavioral Therapy in the Treatment of Rape Victims. In L. Michelson, M. Herson, and S. Turner, eds., *Future Perspectives in Behavioral Therapy.* New York: Plenum.

Veronen, L. J. and D. G. Kilpatrick (1980) Reported Fears of Rape Victims. *Behavioural Modification* 4: 383–96.

Warshaw, R. and M. P. Koss (1988) *I Never Called It Rape.* New York: Harper and Row.

Weis, K. and S. S. Borges (1973) Victimology and Rape: The Case of the Legitimate Victim. *Issues in Criminology* 8: 71–115.

Williams, J. and K. A. Holmes (1981) *The Second Assault.* Westport: Greenwood.

Winkler, C. (1987) Changing Power and Authority in Gender Roles. Ph.D. Thesis, Indiana University, Ann Arbor: Microfilms.

  (1989) Myths About Rape or Prejudices about Rape: DeFamiliarizing the Familiar. Paper Presented at the 99th American Anthropological Association Meetings, Washington, DC.

  (1991) Rape as Social Murder. *Anthropology Today* 1(3): 12–4.

(1992) Comparison of Legal and Physical Rape: The PS Game. Paper Presented at the 91st American Anthropological Association Meetings. San Francisco, CA.
(forthcoming) The Meaning Behind Rape Trauma. In N. Sault, ed., *Many Mirrors: Body Image and Social Relations*. New Brunswick: Rutgers Press.
(n.d.) Raped Once, Raped Twice, and Raped a Third Time (Work in Progress).
Winkler, J. J. (1990) *Constraints of Desire*. New York: Routledge.

12 Words from the Holy People: a case study in cultural phenomenology

Thomas J. Csordas

Embodiment, in the sense I am using it, is a methodological standpoint in which bodily experience is understood to be the existential ground of culture and self, and therefore a valuable starting point for their analysis (Csordas 1990, 1993, 1994). In this chapter I will focus on two issues that must be clarified in advancing a cultural phenomenology that begins with embodiment, or if you will two issues that, unclarified, could become limbs in the embodiment of a straw man. One is the relation between embodiment and biology, and the other is the identification of this phenomenological starting point in preobjective or prereflective experience. I will present each in terms of a problematic quote.

From Martin Heidegger (1977: 204–5) comes a statement of the first problem:

The human body is something essentially other than an animal organism. Nor is the error of biologism overcome by adjoining a soul to the human body, a mind to the soul, and the existentiell to the mind, and then louder than before singing the praises of the mind, only to let everything relapse into "life-experience" . . . Just as little as the essence of man consists in being an animal organism can this insufficient definition of man's essence be overcome or offset by outfitting man with an immortal soul, the power of reason, or the character of a person.

Heidegger implies that the *ad hoc* tacking on of components to an essentially animal body exposes the inadequacy of distinctions among mind/body/soul/ person, and betrays the existential character of the human body as essentially human. Leaving aside the problem of whether Heidegger essentializes the human body to the point of denying the Being of animals (Caputo 1991), does not the requirement to recast our understanding of the body in phenomenological terms thus mistakenly negate the important relationship between biology and culture? In response to this question, from the phenomenological standpoint our answer is that both "biology" and "culture" (or, more specific to the case discussed below, neuropathology and religion) are forms of objectification or representation. Thus a first goal is to suspend our reliance on both – or perhaps to suspend our description between them – in favor of an experiential understanding of being-in-the-world.

The nature of this preobjective being-in-the-world is our second issue, formulated as follows by Maurice Merleau-Ponty:

My body has its world, or understands its world, without having to make use of my "symbolic" or "objectifying" function [1962: 140–1] ... It is as false to place ourselves in society as an object among other objects, as it is to place society within ourselves as an object of thought, and in both cases the mistake lies in treating the social as an object. We must return to the social with which we are in contact by the mere fact of existing, and which we carry about inseparably with us before any objectification (1962: 362).

To deny that the social is an object calls into question the status of "social facts," the existence of which is taken to have been established definitively by Durkheim. Does not the requirement that cultural analysis begin in preobjective experience thus mistakenly presume a pre-social or precultural dimension of human existence? The answer lies in defining the sense in which we carry the social "inseparably with us before any objectification." Accordingly, the second goal of this chapter is to show how cultural meaning is intrinsic to embodied experience on the existential level of being-in-the-world.

The anthropologist addressing issues framed in this way can be distinguished from the philosopher by a simple criterion: the anthropologist is satisfied only by making the argument in terms of empirical data. The data for this chapter's exercise in "fieldwork in philosophy" (Bourdieu 1987) are drawn from the case of a young Navajo man afflicted with a cancer of the brain. I will introduce and contextualize the case in neither biological nor cultural, but in clinical terms. I will then attempt to thread the discussion of being-in-the-world between the two poles of objectification, showing how he brought to bear the symbolic resources of his culture to create meaning for a life plunged into profound existential crisis, and to formulate a life plan consistent with his experience of chronic neurological disease.

Clinical profile

The patient, whom I shall call Dan, was a participant in a larger study of illness experience among Navajo cancer patients (Csordas 1989), carried out with cooperation from Indian Health Service hospitals at Fort Defiance and Tuba City on the Navajo reservation in the southwestern United States. I was able to follow his progress for two years, beginning a year after the onset of his illness in 1985, and ending in 1988, about a year before he succumbed. Dan was 30 years old when I met him, an English-speaking former welder, with an education including two and a half years of college. He came from a relatively acculturated bicultural family; his mother was a schoolteacher and his father a ceremonial leader or "road man" in the Native American Church (peyotist), and one brother was attending college. Dan was divorced and, since the onset of his illness, was closely cared for by his own family.

His diagnosis was Grade II astrocytoma, a left temporal-parietal lobe brain tumor. After the tumor was removed, he received chemotherapy and radiation therapy, and was maintained on medications for control of tremors (legs, right hand, and head) and seizure-type neurological indications. Post-operatively, he experienced chronic headache, hypersensitivity of the operative incision, and occasional olfactory auras and paresthesias. His psychiatric profile was characterized by loneliness, pessimism, self-doubt, poor sleep, low energy, difficulty expressing his thoughts, rigid thought processes, blunted affect, disorganized ideation, rambling speech, preoccupation with mixed strategies for a plan of life, feeling depressed every day, and a sense of irretrievable loss over estrangement from his wife and children; his formal diagnosis included organic personality disturbance and mental deficits.

Dan's rehabilitative status was dominated by the loss and gradual recovery of linguistic ability, accompanied by frustration that "I can't say my thoughts," and that he recognized his relatives but "can't say their names." He reported that something will "hit my mind but I can't say it," and that "I have a hard time trying to mention some words I want to say." His ability with spoken language appeared to return more completely and rapidly than that with written language, and he complained that although his English was steadily improving, the moderate amount of Navajo that he had been able to understand was completely lost. A Test of Non-Verbal Intelligence conducted a year post-operatively resulted in a score of 66, technically indicating persistence of a mild retardation but of ambiguous value based on motivational and cultural factors that could affect test performance.

Dan declined recommended psychotherapy and vocational rehabilitation. His post-operative session of mental testing had left him trembling and with headaches because of the exertion, and he apparently perceived such intervention as contradictory to the doctors' advice not to rush things in returning to normal activity.[1] He expressed some resistance to long trips to an off-reservation university medical center on the grounds that the doctors there "don't do anything" and that he preferred home visits from health-care personnel. In the meantime Dan developed his own rehabilitative strategy for relearning vocabulary upon discovering the existence of word puzzle books, which consist of pages of letters from which one must identify words along vertical, horizontal, or diagonal lines. Completing these puzzles constituted Dan's main activity. Although working too hard on his puzzles sometimes caused him headaches, this activity appeared to provide a self-motivated form of linguistic therapy and cognitive rehabilitation.

By the spring of 1988 Dan's memory and ability to read had returned. Though he had been depressed and remained unsure of himself and his abilities, he claimed once again to be "trying." He had continued working his word puzzles, timing himself to record the increasing rapidity with which he could complete them, and comparing the increasingly neat manner in which he circled the words with his first puzzles, which he said looked like they were done "by a little kid." He acknowledged having "become a perfectionist in dressing and appearance" to such a degree that his sisters teased that the surgeon "must have put extra brain cells in." He had begun once again to talk about having a family, and to joke about "going to Gallup to find a chick." His physical condition at the time was stable.

Cultural phenomenology of language and inspiration

Dan considered himself to be an active person, and expressed frustration at not being able to work. He recounted commenting to his doctors that it was so difficult following their advice not to rush things that "maybe you guys going to have to tie me up . . . [or] go to a special doctor again, what do they do, they put you to sleep, but they talk to you [hypnosis]." Dan's own solution, however, was to "follow the Navajo way," and learn to be "the kind of person that helps people" – a medicine man. This in fact became Dan's preferred strategy for reconstructing his life, focused around his existential struggle for language and expressivity. He indicated that this strategy originated in a direct encounter with the Navajo deities or Holy People, who inspired him with words of prayer:[2]

Yeah. Yeah. See I think, see, I never use to have these help, this kind of help coming, but . . . ah, my life is changing, but right now it is still kind of hurt me . . . sometimes I have some prayers – when I was very small, learning, I prayed in front of my mom and dad or I talked to them – see, some of these [new] words I never knew, I never knew. My mom and my dad just said they never actually heard them [spoken by Navajos]. But then I told them and I said there's stuff I can hear. I said "Somebody waking me up wants me to listen," but then I really have pain in my head so I use to get up. But [when it first happened during Dan's treatment in an off-reservation hospital] we were staying in the motel and my dad was, my dad took care me so we been in the motel all the time, so sometime we go to sleep about 7 then I can't sleep, but when I do go to sleep right about 11 or 12 midnight I feeling [pause] my eyes just open like I'm not sleepy so I say, well, I might as well just turn the tv on, then I go turn the tv on, at that time [shortly post-surgery] too I couldn't understand what tv was, you know, what a tv was or the, the different show, I just look at them, I just sit there like this no laughing about it 'cause I forgot [what it meant] – I don't any more – and from there I just sit there and I said well I guess I can go back to sleep and end up laid back I tried to go – shoot! – just open up my eyes wide again, just like a talking come in, come in, come in and they keep me here for a good hour and a half. So then I get a headache if I don't do no, if I don't do no talking. I can feel it so my dad, I know that he's tired too and he has, I say "Dad please can you sit up." I said, "Some words coming to me I'd like to mention it to you. I want you to tell me if these are right are or if they are wrong. The word that I, that I have." So it's the Navajo way about a long time ago and my dad says, "How can you know because you never even knew that these as you wake up these are put into your brain." And he says then I have to ask someone, I have to talk to either my mom or my dad and ask them each things that sometimes are – I change it that's when they get upset with me ... [confused passage] ... so I switched one of those words and they got caught [i.e., tripped up] my mom and my dad they got caught they could not think why this came out this way so they were trying to think about which one [of his words] was going straight, they were saying that this what they used to do a long time ago. So then as I made it, change it just a little myself I could feel it, just like you're going to throw up, yeah, and the pain real empty again you're just going "Doo! Doo! Doo!" [makes sounds describing a bodily/cognitive sensation] well, and I just sat there and then I just got a hearing that says, "Turn it back, turn it back." So then I just sat there and I, I said, "Mom, Dad," I said, "You have to hear it this way." I said I made a mistake. So then I talked it the way it came to me. Then they could answer it so they got caught on the spot.

This episode occurred relatively early in Dan's post-operative recovery, when his cognitive abilities were so impaired that he could not even comprehend television programs. It comprised a lengthy auditory experience, followed by a compulsion to talk that relieved his intense headache pain and left him with a "happy and a good feeling." For Dan this spiritual help received from the Holy People was different from the way he learned to pray as a child. The help does not come as an answer to prayer – it is the ability to pray itself that constitutes help for Dan, and his family concurred that indeed he was never before able to pray like he does now. He also reported that when he asked his father, himself a ritual leader, why he was speaking thus, his father responded that "Someday you are going to be a very powerful person to help people." Some of his younger relatives, who he has already begun to encourage with his new-found wisdom, have said "Uncle Dan, we kind of know that you are going to be a medicine man."

Nevertheless, his condition made the cultural validation of his experience, which was critical to Dan, somewhat problematic. During the struggle for fluency that characterized his recovery, Dan's parents and others had considerable difficulty in understanding him. This difficulty was compounded of severe linguistic impairment and of the understanding that the utterance was a direct revelation of a new, contemporary synthesis of traditional Navajo philosophy in a young person who had never before known such things. Dan indicated that his own efforts to correct his speech were ineffective, and that only when he consented to "turn it back" and let his speech come out as it was inspired did they begin to understand.

His parents also told him that he must speak before the elders in the peyote meeting for their confirmation. He did this during a series of four peyote meetings held for his post-surgical recovery. These meetings last an entire night, during which participants ingest peyote, sing, and take turns uttering often lengthy, spontaneous, inspired prayers (Aberle 1982; La Barre 1969). In this setting each of the ritual officiants, the patients, and frequently other elders pray formally, and there are intervals in which quiet conversation is permitted. These intervals typically include encouragement and exhortation of the patient. Dan reported beginning his speech with the acknowledgement that what he was going to say might not be understandable to his elders because he was a younger person and his words would be "brand new," and receiving their permission to speak on the grounds that "if you want to help people [with] what you are saying, we have to listen to you." He said, "I made my heart and prayed to god, first, ask him, is this what I want for the rest of my life [to be a medicine man]?"

He also told them of his ambition to spend four years as a cowboy (i.e., a professional bareback rider) prior to becoming a medicine man. Dan's anomalous desire to be a rodeo cowboy, which persisted throughout the time I knew him, is very likely the element that led the consulting psychiatrist to report "mixed strategies" in his thinking. In Dan's narrative it appears as a consistent part of his plan, though quite unrealistic given his physical condition. He describes the hoped-for cowboy career as lasting for a specific period of four years prior to becoming a medicine man, indicating by the use of the sacred number four that it would constitute a preparatory period. In addition to satisfying a personal wish, it would prove that he was physically competent, making money, and would perhaps help to establish a reputation that could be transferred to work as a medicine man.

Although Aberle (1982) points out that, in contrast to traditional Navajo ritual, it is virtually impossible to make a mistake in peyote prayer, from Dan's narrative it was evident that his speech in the peyote meeting provoked criticism. As he said, "Sometimes they get after you, like if you talk wrong or if you speak and it's wrong, or if your thinking is wrong. That's when they get after you, they tell you." Given his linguistic disability, his innovative or idiosyncratic message, and his proposal to become a medicine man via a rodeo cowboy career, the reaction to his speech was, not surprisingly, somewhat mixed. Dan acknowledged that some people accused him of being "wrong" and "off" in what he was saying, even interrupting him in violation of the ritual protocol of peyote meetings, to the point of making him cry. This reaction appears to have been mediated by the leader of the ceremony, however, who both accepted the legitimacy of what Dan said and acknowledged that the problem is "just the way you talk." Noting that Dan's father would be getting old and that the Navajo deities or Holy People appeared to be indicating that Dan could eventually replace him, the leader concluded, according to Dan's report, "'Some day you are going to be a helpful person,' that's how they said it to me ... So from there I'll always be helpful."

Six months later, in another account of the same significant incident, Dan acknowledged making mistakes in what he was saying that upset people, and a certain arrogance in appearing to "force" his youthful inspired message on the others. He said that in trying to explain his thoughts he had to "change it back and then there's no hard feelings," but in his view, Dan eventually won over the elders. There appear to be three aspects to the validation of his claim.

First, in crediting his healing to spiritual help obtained in earlier peyote meetings, he claimed that the divine inspiration that was a consequence of the healing should be listened to by other peyotists. He reinforced this claim by stating that he could speak in this manner even though his "brain was cut out," and that others with similar problems do not recover their speech at all: the very fact that he could talk at all validated his words. Elsewhere in our interviews he stated that he never used to talk as he does now, evidently referring not only to the content of his speech but to the fact that before his illness he was rather taciturn.

Second, invoking a bodily criterion of validity, the other participants acknowledged that Dan would have become ill if his prayer had been incorrect. It is well known that peyote ingestion can cause severe vomiting. One interpretation is that the Holy People cause this suffering as a punishment for wrong thinking or speaking. Since Dan was unaffected by the peyote his words were finally understood as incurring divine approval.

Third, Dan argued that part of the reason for his words being misunderstood was that they were addressed to the contemporary situation and problems of the younger Navajos. While acknowledging that the older people can help troubled youth with peyote meetings, he emphasized that there are now many more people in the world and that things are different than they were twenty-five years ago. Old people know only the reservation, but younger ones have traveled and have even been to colleges, and so are being inspired with different kinds of prayer. According to Dan, the participants acknowledged that his words could help them better understand their own grandchildren. He claimed that some were moved to tears by his words, a reaction in conformity with the common occurrence of weeping in heartfelt response when one is moved by the sincerity of the speaker during peyote prayers (La Barre 1969: 50, Aberle 1982: 156).

The experience of language again played a remarkable and poignant role with respect to the theme that his message was intended for Navajo youth. Even though English was his primary language and by both his account and that of his mother his knowledge of Navajo was quite minimal, he attributed a great deal of meaning to the apparently permanent loss of linguistic facility in Navajo while having recovered the ability to speak in English:

So I said "I guess that's what you call I'm learning to be a Navajo person, I guess," [pause] but then I said the reasons why is that maybe I am going to be the helper because there is thousands of people, I think, I think that's why I lost my Navajo talk, because there's thousands, thousands of people that went to school there [in] colleges, they don't understand right but then sometime they want help but then they can't understand the Navajo so they get upset. They're new people, they just do the whole work for them in the Navajo way and they can't really stand up, they don't know what's being said so [they] kind of get upset about it so they don't know which way to go again.

He concluded that he lost his Navajo because the Holy People wanted him to address his message in English to the young people who wanted help from traditional Navajo ceremonies, but who became frustrated because they could not understand ceremonial proceedings. Said Dan, "maybe I can give them that."

That Dan attributed a great deal of his recovery to the divine help afforded by peyote is evident in the following passage from our second interview. The context of the passage is my query about his response to peyote during the meetings held for his benefit, a query intended to identify any experiential interaction between effects of the psychoactive substance and his neurological condition:

[The peyote] didn't really bother me. It just [pause] the way, what it was, it just brought me to where that I can use the mind to think ... But if I didn't [pause] I would have nothing. See, that medicine is what [pause] they [the Holy People] give us, it is what we use. That's the ones that I eat it [pause] and it goes throughout my mind. It brings things to me [pause] for new types are the ones are sent [pause] and that's when it goes into me [pause] I start doing my talking. Before I never use to talk this way but now there's different word that all comes out [pause] especially when I have prayers [pause] I ask questions and then just before I go to sleep, sometimes I [pause] I have a prayer [pause] and I then I wonder that's why I always say "Heavenly Father, I wonder." I'm a wondering person [pause] as young as I am, for some reason I want to be a helper of people, young people my age, some of them may be older than me. So I want to be that type of person ... So then sometime I sleep [pause] and when I sleep here [in this house?] everything's close to me. A feeling comes into me, some big words, everything that I never thought about so when I get up in the morning I speak to my Mom and my Dad [pause] especially to my Dad [the peyote road man] more, and I ask him are those types of word, are they right or are they wrong? ... [I ask: Can you tell what some of those words are?] It's like [pause] I say, "Heavenly Father," I said, "At this time my life is going through a very hardship but the type of word maybe I have not spoken heavenly father maybe that's what holding me back. I want you to teach me heavenly father [pause] my life has [pause] no [sigh] happiness within me. But then from type [unclear] in me father, is the type that would, is more specially the Navajo people [pause] to put them together, don't force them, easy, don't force them." See these [kinds of words] started coming out, but I never use to do, because people they sit [in peyote meetings] and they look [pause] they just start praying real low. Then they speak to me in the morning [at the end of the meeting] they say "You coming Dan," [pause] they say, "You never use to pray that way." I said, "That's what I've been going through and it makes me feel a lot happier a lot better ... that my mind is actually coming.

The ability to think and speak "straight" are given great priority in Native American Church practice (Aberle 1982; La Barre 1969) and in general among Navajos (Witherspoon 1977), and Dan explicitly attributed his ability to do both to peyote as it "goes throughout my mind" and "brings things to me" so that "I start doing my talking." The brief sample prayer that Dan spoke for me also appears quite appropriate in the peyotist context. "Heavenly Father" is a common formula of address in Navajo peyote prayers (Aberle 1982: 153).

Dan's notion that there is a "word maybe I have not spoken" that is "holding me back" is of particular interest, for if it is true that in traditional Navajo religion there is a concern for absolute accuracy of utterance within a fixed liturgical canon, there is an equivalent concern in peyotism with the spontaneously creative utterance of the ritual *mot juste*. This takes on a particular significance in the case of a person afflicted by anomia, for whom rehabilitation and inspiration are synthesized, and the ability to utter exactly the right words signify both personal healing and the ability to help others. The salience of this synthesis is borne out in the very next statement, in which Dan prays that the Navajo people will be able to come to terms with the future gently and asks the divine "don't force them" – precisely the advice given to Dan by his physicians concerning rehabilitation.

Other statements in our interviews indicate that Dan's message was in conformity with traditional moral themes of Navajo peyotism, including concern for students away from the reservation, the importance of education in the contemporary world, relations with white people, and identification with the United States as well as with humanity as a whole (Aberle 1982: 154, 156). At one point Dan gave a lengthy discourse about the traditional concept of the rising and falling life cycle as symbolized by the crescent-shaped peyote altar, lamenting how old people were neglected by contemporary youth and left in nursing homes. Another instance highlights global moral concerns:

this earth is very small and it's packing and packing throughout the world, and ... to [avoid] having a war and have a big what do you call that, the Russian people ... For some reason it hit me and made me speed up [sigh] trying to learn and trying to start praying, praying the good way, to hold down [the war], so we don't want to go over there and start to fight again, have to fight. I said this is our land, white people that came over here . . . but since we are all mixed together, forcing us Indians to help, lot of them are going to get hurt, a lot of them aren't going to come home . . . And then sure enough [pause] what happened with them . . . They have a big blow out. In Russia. Poison. Some of it went all the way around. Already hit me – I already knew it. So then we had to ask the United States to send some people over there that are smart so that they can plug that thing up. That's what happened. So then they didn't want to argue any more they were just happy with each other . . . I said that's just like the Navajo way, you know. They trying to, they are learning ahead of time, I guess that's what it is. They try to warn people ahead of time. And sometime they can't understand, you know. It gets too close. But then you always, you just mention, you can't just force them. You can just help them.

In this excerpt Dan recounts a prophetic experience that both encouraged his spiritual aspirations and impressed upon him the moral urgency of global concerns. The experience was of knowing about the Russian nuclear disaster at Chernobyl before it occurred. The broadening circle of danger that threatens the Navajos by virtue of their cooperation with the United States in any conflict with the (former) Soviets, and that likewise identifies American goodwill in offering friendly technical assistance to the Soviets with the "Navajo way" of harmony, is consonant with the "prayer circle" in which the prayers of Navajo peyotists generate "gradual spread of the blessings from the immediate group of those present to the whole world" (Aberle 1982: 153).

This cosmological implication caps the existential analysis of language for Dan. Being a real Navajo person meant having the actuality of language as a mode of engagement in the world, having the project of becoming a medicine man was the rationale that grounded the return of language, and peyotist spirituality defined the moral horizon of his discourse as a global horizon. His struggle was not for *langue*, an abstract cognitive or representational ability, but for *parole*, the ability to produce coherent, socially and morally relevant utterance.

Neurology and cultural phenomenology

Let us remind ourselves that despite Heidegger's rejection of the notion of the body as an "animal organism," the phenomenology of embodiment has engaged in theoretical and empirical dialogue with biology, specifically in Merleau-Ponty's (1962) analysis of patients with neurological lesions. To this point my exposition has been a hermeneutic of Dan's struggle for meaningful utterance, a hermeneutic that shows how he thematized and objectified his embodied experience of language into a life plan in conformity with cultural and religious meaning. We must now enter this analysis into dialogue with the considerable literature on experiential and behavioral consequences of temporal lobe lesions. Although this literature focuses almost entirely on patients with epilepsy rather than the relatively more rare brain tumor, the persistence of a secondary seizure disorder following resection of a left temporal-parietal astrocytoma warrants an examination of left temporal-lobe epilepsy for insight into several formal features of Dan's post-surgical experience.

Two broad groups of behavioral syndromes are of possible relevance. First are the so-called schizophrenia-like psychoses of epilepsy, which are episodic in character and clinically similar to atypical psychoses, and in which the majority of patients have either bilateral or left temporal-lobe involvement (Slater et al. 1963; Tucker et al. 1986). Second is the so-called interictal behavior syndrome of temporal-lobe epilepsy, which describes a complex of behaviors persisting through the period between explicit seizure or ictal activity and indicating enduring personality changes due to the illness (Waxman and Geschwind 1975). A critical feature observed in notable proportions of patients in both categories bears the clinical label "hyperreligiosity." The first documented sample of schizophrenia-like psychoses included 26 (of 69) patients who exhibited such religiosity, of whom only eight were religious prior to the onset of their illness, and of whom six experienced profound and sometimes repeated episodes of religious conversion (Dewhurst and Beard 1970). In a more recent study of patients originally admitted for behavior disturbance and secondarily diagnosed with seizure disorders, 5 of 20 patients exhibited hyperreligiosity (Tucker et al. 1986). In the series of cases cited to define the interictal behavior syndrome, 6 of 9 patients exhibited some degree of religiosity (Waxman and Geschwind 1974, 1975).

It quite reasonably has been hypothesized that in spite of the overlap with respect to features such as hyperreligiosity, there is a degree of distinction among classic ictal phenomena proper, the episodic clusters of affective features that define the epileptic psychoses, and the nonremitting interictal behavior syndromes (Tucker et al. 1986). For Dan, it appears that actual seizure activity was largely controlled by medication, and there is no evidence justifying the label of epileptic psychosis. However, the following features of the interictal behavior syndrome suggest its relevance in Dan's case:

Deeply held ethical convictions and a profound sense of right and wrong ... interest in global issues such as national or international politics ... striking preoccupation with detail, especially as concerns moral or ethical issues ... preoccupation with detail and clarity [of thinking] and a profound sense of righteousness ... speech often appears circumstantial because of the tendency of these patients to digress along secondary and even tertiary themes ... deepening of emotional response in the presence of relatively preserved intellectual function. (Waxman and Geschwind 1975: 1584)

A second set of authors describes these changes not as a behavioral syndrome but as a personality syndrome that includes hypermoralism, religious ideas, an unusually reflective cognitive style, tendency to label emotionally evocative stimuli in atypical ways, circumstantiality, humorlessness, difficulty producing the names of objects, and an obsessive, sober, and ponderous style probably related to an effort to compensate and make sense of a world rendered exotic by language dysfunction and confusion about temporal ordering of cause and effect relationships (Brandt et al. 1984). These are all features that we have encountered in some form in Dan's case, and their relevance is supported by the observation that they occur primarily in patients with left temporal-lobe epilepsy, suggesting their origin in disruption of left temporal-lobe mechanisms (Brandt et al. 1984).[3]

We must give special attention to disturbances of language in temporal-lobe epilepsy, since language plays a critical role both in Dan's frustration over his anomia and in his divinely inspired ability or compulsion to pray. Two features of the interictal behavior syndrome are relevant: hypergraphia or the tendency to write compulsively and often repetitively, and verbosity or loquaciousness, the tendency to be overly talkative, rambling, and circumstantial in speech. Perhaps due to a combination of illness-related cognitive deficit and the persistent trembling of a right-handed person with a left-hemisphere lesion, Dan claimed to be virtually unable to write at all, and unless one considers his dedication to word puzzles as a form of hypergraphia this feature must be judged to be absent. On the other hand, verbosity is evident in both his interview transcripts and his ability to generate lengthy peyote prayers. Accordingly, it is consistent with our argument that while hypergraphia is typically a characteristic of right temporal-lobe epilepsy (Roberts et al. 1982), verbosity is associated with left temporal-lobe epilepsy (Mayeux et al. 1980; Hoeppner et al. 1987).

An important hypothesis for our purposes stems from the observation that left temporal-lobe epileptics perform worse than other epileptics on confrontation naming tests and related dimensions of verbal functioning (Mayeux et al. 1980). The researchers suggest a location of the lesion in the inferior left temporal-lobe, and propose that the integration of sensory data requisite to naming occurs in the parietal lobe. This would affirm the importance of the temporal-parietal site of Dan's tumor in his acknowledged anomia. However, the critical point of their argument is that loquaciousness is quite likely an adaptive strategy compensating for a patient's inability to find the right word, and neither a direct effect of the lesion nor an epileptic "personality trait" as is commonly assumed. Jenkins (1991) has observed the cultural predisposition in American psychology and psychiatry to explain behavior by personality attribution, a predisposition which in this case might obscure the importance of intentionality and reconstitutive self process. I would suggest that this hypothesis is supported by a review of Dan's interviews, where much of the apparent ponderousness may be attributable to the frequent false starts and rephrasings that indicate a search for the appropriate word.

Dan's desire to become a medicine man is closely bound up with his search for the "word maybe I have not spoken heavenly father maybe that's what holding me back" and his goal of teaching and helping as a medicine man through the utterance of lengthy peyote prayers. Compare this personal scenario with the case described by Waxman and Geschwind of a 34-year-old white man who had undergone several craniotomies for a left temporal-parietal abscess, precisely the location of Dan's lesion:

Three years after surgery ... his hospital records noted that "he now expresses interest in religion and possibly in becoming a minister, and hopes to increase his 'use of good words' towards that end ..." He subsequently at his own suggestion delivered several sermons at his church. He spontaneously delivered typed copies of the texts of these sermons to his physician. The sermons concerned highly moral issues, which were dealt with in highly circumstantial and meticulous detail. (Waxman and Geschwind 1974: 633).

Dan's anomia and this patient's aphasia – or, more directly to our point, the shared features of their recoveries – could certainly be objectified in neurological terms, but the inability to find a word bears a preobjective existential significance that also becomes objectified or thematized primarily in religious terms. Stated somewhat differently, the remarkable similarity of these cases is very likely due to the similar site of the neurological lesion, but while the lesion comfortably accounts for the linguistic disorder, it does not adequately account for elaboration of the moral significance of finding the right word. This leads directly back to bodily existence and "the social that we carry around with us prior to any objectification." In Dan's case we can discern this existential ground in two ways: with respect to bodily schemas that organize interpretations of etiology, prodromal experience, and post-surgical sequelae; and with respect to the existential status of language itself.

Dan's understanding of how his cancer occurred followed the traditional Navajo etiological schema[4] of contamination by lightning (cf. Csordas 1989). Two points must be made in order to understand the preobjective cultural reality of this attribution. First, for Navajos lightning is not a simple natural phenomenon, but includes all kinds of radiant energy such as radioactivity and emanations from microwave ovens. Dan cited a boyhood exposure to natural lightning while herding sheep as the origin of his tumor, but also cited the adult exposure to the flames and fumes of his welding torch as a secondary exposure to lightning that precipitated the onset of his current illness. Second, lightning is not an abstract spiritual phenomenon that acts by mystical contagion devoid of cultural phenomenology. The prototype of lightning contamination is direct physical encounter in which lightning strikes so near that the person sees a bluish flash and inhales acrid electrical fumes (*il hodiitl'iizh*; cf. Young and Morgan 1987: 453). The preobjective grounding of this contamination schema in bodily experience accounts for the assimilation of natural lightning and the lightning produced by the welder's torch.

The schema that preobjectively ordered Dan's experience of the prodromal period was that of the ritual importance of "fourness" among Navajos, as in four cardinal directions, four sacred mountains, the performance of ceremonies in groups of four, etc. Highly salient in Dan's personal synthesis are three discrete events that occurred within several months of each other and culminated in his hospitalization and diagnosis. The first occurred on a trip with some co-workers to a large off-reservation city, when he fell unconscious from the second-story balcony of a motel. The second occurred while he was alone at home with his father, when Dan started shaking uncontrollably while sitting on the couch preparing to read a document that his father could not understand. The final episode occurred while Dan was at lunch with co-workers, when he collapsed in a restaurant and was taken to the hospital. For him the possibility of a life-culminating fourth seizure was deeply implicated in his daily life and in his plan to become a medicine man. In other words, the social that Dan carried about with him as a feature of his embodied existence was the profound polysemy of the sacred number "four."

Dan did in fact experience something close to this fourth seizure in the post-surgical period of struggle to regain language and expressivity. He was sitting in his room listening to the radio when it began to emit the high-pitched test tone of the Emergency Broadcast System. Dan reported, "I had a feeling that something went through me, I was just in here going like this [shaking]." The cultural basis for this preobjective experience is the schema of a foreign object shot through one's body by a witch. This schema was explicitly elaborated by a medicine man who determined that there was a sharp fragment of bone inserted or shot into Dan's brain by a witch, and who removed it by appropriate ritual means. Dan felt that the bone had been inserted into his head so that he would "not to be ever able to have that kind of thought, not to be able to speak again."

The existential status of language for Navajos was captured by Reichard (1944), who called attention to their self-consciousness with respect to words and word combinations, and subtitled her description of Navajo prayer "The Compulsive Word." Reichard's phrase has a double meaning in that prayer compels the deities to respond, and that efficacy requires a compulsive attention to correctness of utterance. Dan's case adds a third sense, for it is he who is compelled to speak by the Holy People. The implication is that because they inspired the words the deities will be more readily compelled by them, but with the paradox that the speaker must struggle to give them coherent and lucid form. This layering of compulsion upon compulsion conforms to a schema that undergirds all of Navajo thought about healing:

According to Navajo belief that which harms a person is the only thing that can undo the harm. The evil is therefore invoked and brought under control by ritualistic compulsion. Because control has been exerted and the evil has yielded to compulsion it has become good for the person in whose behalf it has been compelled. In this way evil becomes good, but the change is calculated on the basis of specific results. (ibid.: 5)

For Dan the evil of illness is transformed into the medium by which he can become a "helping person," that is, language which is at once compulsive and compelling.

Dan's experience cannot be understood outside the intimate and efficacious performative nexus among knowledge, thought, and speech. Reichard notes that unspoken thought and spoken word share the same compulsive potential (ibid.: 9–19), and Witherspoon (1977) has shown that for Navajos, thought is the "inner form" of language, and knowledge the "inner form" of thought. All three are powerful, in that all knowledge is inherently the power to transform or restore, all thoughts have the form of self-fulfilling prophecies, and all utterance exerts a constitutive influence on the shape of reality: "This world was transformed from knowledge, organized in thought, patterned in language, and realized in speech ... In the Navajo view of the world, language is not a mirror of reality; reality is a mirror of language" (ibid.: 34). In this light, the gift of speech, experienced in the modality of divine "otherness" as a spontaneous gift from the Holy People, is coterminous with the intuition of spiritual knowledge that should be used to help others. Also coterminous are Dan's struggles for a post-operative return to fluency and for the ability to utter a spiritually relevant message. Because of the existential status accorded to speech, the fact *that* Dan was trying to speak bore implications for *what* he was trying to speak.

With this glimpse at the habitus within which Dan's struggle for language took place we can tack once again back to an issue raised by a clinical study, this one of non-Navajo verbose epileptic patients with left foci and complex partial seizures. This group of patients was relatively older than others studied by the same researchers, and showed no notable anomia and no poorer performance on verbal tasks than other epileptic patients. However, their performance on a story elicitation task was characterized by "nonessential and at times peculiar details" (Hoeppner et al. 1987). Of interest to us are the three alternative hypotheses offered by the researchers to account for this verbosity: patients are either driven internally to talk, notice things not noticed by others, or give particular significance to things others see as irrelevant. For the authors these hypotheses indicate a relation between verbosity and hallucination. In particular, they suggest that there may be a continuum between noticing or commenting on details trivial to others and hallucination. Reasoning from this idea, verbosity may be a "subclinical manifestation of a complex phenomenon, involving speech, perception, and affect, which in more severe form appears as psychopathology" (ibid.: 40).

Dan explicitly acknowledged being internally driven to talk, with refusal to do so resulting in a headache. We can infer from his narrative that at first his experience of hearing the Holy People was independent of the words which he then repeated in order to verify their validity. It further appears that over time the experience became a direct non-auditory inspiration to talk. From the initial account of hearing words shortly after his surgery, his description of the inspiration changes to having the words given to him or put into him, and eventually to simply having the peyote spirit enter him and that's when "I start doing my talking." Thus beginning with what are probably best described as auditory hallucinations, which are clinically more common than visual ones in temporal-lobe epilepsy, it appears that there was a phenomenological fusion of what Dan heard with what Dan said. For Dan, then, the adoption of the goal to pray as a medicine man served as a self-orientation that allowed for the domestication of auditory hallucination into intentional (if inspired) utterance, for a devout attention to details ignored by others, and for a search for the deeper significance of the apparently irrelevant. Here we can go no further, for we are at the very frontier of the cultural phenomenology of language.

The question of hallucination leads directly, however, to the neurological and psychotropic effects of peyote ingestion. One might assume that the development of "verbose" prayer in peyote meetings was stimulated by a drug whose neurological effects parallel those of temporal-lobe anomalies, but this would be a reductionistic slight to firmly established canons of American Indian oratory that have no discernible link to specific brain structures. Nevertheless La Barre (1969: 139–43), in his careful study of the peyote cult, lists effects of the cactus including talkativeness, rambling, tremulousness, rapid flow of ideas, shifts in attention with maintenance of clarity, incoordination of written language. Auditory hallucinations stimulated by peyote primarily modify the quality of sounds, although La Barre would perhaps accept the possibility of voice audition such as Dan experienced. Although peyote is known to produce convulsions in animals (one of its constituent alkaloids has properties similar to strychnine), it is not known to produce seizures in humans, and localization of its action remains unspecified. The important phenomenological work of Kluver (1966) suggests the involvement of the vestibular system and that the effects of parieto-occipital lesions exhibit some formal similarities with those of peyote, but the emphasis of his study was on visual effects, none of which play a role in the case under consideration.

The single extant study on mescaline administered to epileptic patients showed that the principal effect on 8 of 12 was drowsiness, lethargy, and apathy or sleep, concordant with the somewhat narcotic properties of that constituent alkaloid of peyote when taken by itself (Denber 1955). A more significant finding of this study was that, although there was no correlation between brain wave (EEG) and clinical results, there was a decrease or disappearance of delta waves, and a disappearance of spike-wave patterns and slow waves lasting for hours. The author observes that problem-solving activities also cause a disappearance of spike-wave patterns, but only for several minutes. Here we might suggest that as a form of problem-solving, Dan's word puzzle activity may have had some effect in suppressing seizure activity, conceivably reinforcing the effects of peyote. In addition, to the extent that inspired prayer became part of Dan's intentional synthesis of a life project it could be understood both as a kind of moral problem-solving activity, and as an action that reinvoked the peyote experience in the manner typically referred to as a psychedelic "flashback." In this light the boundaries between the effects of neurological lesion and cultural meaning dissolve, revealing in the space between a glimpse of the preobjective indeterminacy of lived, embodied experience.[5]

Finally, brief mention is made of peyote in Levy et al.'s (1987) study of seizure disorders among the Navajo. Interestingly, one patient was dropped from the study because her family was peyotist and had not diagnosed her, implying either that Navajo peyotists do not recognize epilepsy as a disorder or regard it as a problem to be treated exclusively with peyote medicine. In fact, the authors noted that "some participants in peyote meetings would hold one arm to keep it from shaking. This was taken as a sign that some malign influence was at work and that the individual in question needed special prayers" (ibid.: 104).[6] I do not know whether Dan's arm trembled during his peyote meetings, but the general persistence of such tremors in his right arm may be closely connected to the medicine man's diagnosis of malign influence in the form of witchcraft.

Meaning and lesion in bodily existence

The connection between epilepsy and religion has long been a *bête noir* for anthropology. For decades it was assumed that shamans and religious healers were either epileptic or schizophrenic, and the struggle to normalize them in the literature was a struggle of cultural relativism against biological reductionism, against the pathologization of religious experience, and on behalf of sensibility and meaning where there appeared to be only bizarre and irrational behavior. Anthropology fought against the logical error that the religiosity of the epileptic necessarily implied epilepsy in the religious. Nevertheless, as late as 1970 Dewhurst and Beard wrote an article describing the cases of several temporal-lobe epileptics who had undergone religious conversion, and juxtaposed them to a series of "possible epileptic mystics" from the Western tradition, including among others St. Paul, Theresa of Avila, Theresa of Lisieux, and Joseph Smith. One might suspect the presence of a rhetorical agenda in a study of religion that begins with the religiosity of explicit medical patients. Such an agenda is indeed evident in the following passage:

Leuba's (1896) so-called psychological analysis of the mental state immediately preceding conversion includes such theological categories as a "sense of sin and self-surrender"; there is no mention of temporal lobe epilepsy. (Dewhurst and Beard 1970: 504)

The authors allow no analytic space between the theological and neurological, despite William James's (1961 [1902]) extensive use of Leuba's work that certainly acknowledged it as "psychological." Dewhurst and Beard (1970) clearly disdain the study of a "mental state" in favor of determination of a "neurological state." Yet even they acknowledge that the eminent neurologist Hughlings Jackson did not attribute religious conversion directly to epileptic discharge, but argued that it was facilitated by the alteration in level of consciousness, the increased excitability of lower nervous centers, and religious background.

A more recent series of papers by Persinger (1983, 1984a, 1984b; Persinger and Vaillant 1985) attempts to validate a more subtle hypothesis about the relation between religion and temporal-lobe function. He argues that "religious and mystical experiences are *normal* consequences of spontaneous biogenic stimulation of temporal lobe structures" (Persinger 1983: 1255). Thus religious, paranormal, and ictal phenomena are distinct, except that they all originate in electrical activity of the temporal lobe. The hypothesis is predicated on the post-stimulation instability of temporal-lobe structures, susceptibility to transient vasospasms, and the tendency of cellular membranes to fuse. The latter contributes to unusual mixtures of cell ensembles especially sensitive to a variety of factors including tumorigenesis. Persinger proposes a construct of Temporal Lobe Transient (TLT), defined as learned electrical microseizures provoked by precipitating stimuli and followed by anxiety reduction, and his work attempts to demonstrate that persons who report religious and paranormal experiences also report signs of temporal-lobe activity. Whether or not one is convinced by these results, it is important to note that Persinger equivocates between regarding religious experience as normal and as an evolutionary liability. Even more salient is his equivocation between neurological reductionism (biogenic origin) and statements that people are predisposed to TLTs by cultural practices or "racial-cultural" interactions, and that "metaphorical language is the most profuse precursor to TLTs" (ibid.: 1260).

What is absent from such accounts is the analysis of the embodied, speaking person taking up an existential position in the world. Without this we risk a battle of causal arrows flying in both directions from neurology and culture, with no analytic space between. By the same token, eliminating the preoccupation with causality is part of phenomenology's radicality in opening up the space of being-in-the-world to existential analysis. Following this line, our argument in the case of Dan suggests that it may make sense to consider verbosity not as a personality trait grounded in neuroanatomical change but as an adaptive strategy that spontaneously emerges from a preobjective bodily synthesis. Likewise, it suggests that the patient's search for words is thematized as religious not because religious experience is reducible to a neurological discharge but because it is a strategy of the self in need of a powerful idiom for orientation in the world.

Here we rejoin Heidegger and Merleau-Ponty; one condemning the error of biologism that consists in merely adding a mind or soul to the human body considered as an animal organism, the other condemning the error of treating the social as an object instead of recognizing that our bodies carry the social about inseparably with us before any objectification. The errors are symmetrical: in one instance biology is treated as objective (the biological substrate), in the other the social is treated as objective (the domain of social facts), and in both instances the body is diminished and the preobjective bodily synthesis is missed. For biologism the body is the mute, objective biological substrate upon which meaning is superimposed. For sociologism the body is a blank slate upon which meaning is inscribed, a physical token to be moved about in a pre-structured symbolic environment, or the raw material from which natural symbols can be generated for social discourse.

Approaching the same problem from feminist theory, Haraway (1991) argues that the body is not an objective entity because biology itself is situationally determined: "the 'body' is an agent, not a resource. Difference is theorized *biologically* as situational, not intrinsic, at every level from gene to foraging pattern, thereby fundamentally changing the biological politics of the body" (ibid.: 200). With the recognition that difference is not merely a cultural overlay on a biological substrate, our argument goes beyond the pedestrian assertion that culture and biology mutually determine the experience of illness, and toward a description of the phenomenological ground of both biology and culture. The struggle for correct expressive utterance has a global and corporal existential significance, whether the ones who struggle are Indian or Anglo, and whether or not they are neurologically afflicted.[7] Only late in the process of analytic objectification can we say that the features of an interictal behavioral syndrome are formally dependent on the nature of the neurological lesion, and that the options to become a medicine man or minister, to speak in the genres of peyote prayer or sermon, and to search for the correct word that will bring healing for oneself and help to humankind are formally dependent on culture. The problems posed by Heidegger and Merleau-Ponty thus point us not toward the end but toward the starting point of analysis. That starting point is a cultural phenomenology of embodied experience that allows us to question the difference between biology and culture, thereby transforming our understanding of both.

ACKNOWLEDGEMENTS

The research reported on in this paper was supported by grants from the National Center for American Indian and Native Alaskan Mental Health Research, the Milton Fund of Harvard Medical School, and the Arnold Center for Pain Research and Treatment of the New England Deaconess Hospital.

NOTES

1 This raises the issue of cultural assumptions about rehabilitation and recovery, a topic as yet very inadequately examined. It is possible that the Anglo-American assumption that some form of formal rehabilitation should begin as early as possible is here contradicted by a Navajo assumption that one should wait until one is rehabilitated, or until one's capacities "come back" before entering any formal retraining, even if that retraining is itself oriented toward not over-exerting the patient.

2 The transcriptions of Dan's words give evidence of continued illness-related linguistic disability, and for that reason they have not been edited to increase their fluency.

3 Most authors report a latency of several years between onset of epilepsy and emergence of the behavioral syndrome; Dan's case study was completed between twelve and thirty months after his surgery.

4 My use of the term schema is loosely related to the notion of image schema put forward by Mark Johnson (1987). In this context I use the term out of convenience rather than conviction, for Johnson's approach is essentially cognitive rather than phenomenological. However, not wishing to open a large area of theoretical debate, I would suggest that, appropriately elaborated as part of a cultural phenomenology, the notion of "habit" would perform an equivalent theoretical role and for our purposes be more apt than that of "schema."

5 On the concept of indeterminacy in embodiment theory, see Csordas (1993).

6 This is evidently different from the traditional Navajo interpretation of such trembling as a sign that the person is a candidate for initiation as a diagnostician or "handtrembler."

7 As Ortner (1974) has argued with respect to the universality of male dominance and the symbolic assimilation of males to culture and females to nature, the existence of similar strategies across cultures and religions does not place such strategies outside culture. Even less does it place them outside "being-in-the-world."

REFERENCES

Aberle, David (1982) *The Peyote Religion among the Navajo*. Chicago: University of Chicago Press.

Bourdieu, Pierre (1987) *Choses Dites*. Paris: Editions de Minuit.

Brandt, Jason, Larry Seidman, and Deborah Kohl (1984) Personality Characteristics of Epileptic Patients: Controlled Study of Generalized and Temporal Lobe Cases. *Journal of Clinical and Experimental Neuropsychology* 7: 25–38.

Caputo, John D. (1991) Incarnation and Essentialization. *Philosophy Today* (Spring): 32–42.

Csordas, Thomas J. (1989) The Sore That Does Not Heal; Cause and Concept in the Navajo Experience of Cancer. *Journal of Anthropological Research* 45: 457–85.

(1990) Embodiment as a Paradigm for Anthropology. *Ethos* 18: 5–47.

(1993) Somatic Modes of Attention. *Cultural Anthropology* 8: 135–56.

(1994) *The Sacred Self: A Cultural Phenomenology of Charismatic Healing*. Berkeley: University of California Press.

Denber, Herman C. B. (1955) Studies on Mescaline III. Action in Epileptics: Clinical Observations and Effects on Brain Wave Patterns. *Psychiatric Quarterly* 29: 433–8.

Dewhurst, Kenneth and A. W. Beard (1970) Sudden Religious Conversions in Temporal Lobe Epilepsy. *British Journal of Psychiatry* 117: 497–507.

Haraway, Donna (1991) *Simians, Cyborgs, and Women: The Reinvention of Nature*. New York: Routledge.

Heidegger, Martin (1977) Letter on Humanism. In *Basic Writings*, David Farrell Krell, ed. (with Introductions). New York: Harper and Row, p. 230.

Hoeppner, Jo-Ann, David Garron, Robert Wilson, and Margaret Koch-Weiser (1987) Epilepsy and Verbosity. *Epilepsia* 28: 35–40.

James, William (1961) [1902] *The Varieties of Religious Experience*. New York: Macmillan.

Jenkins, Janis (1991) Expressed Emotion and Schizophrenia. *Ethos* 19: 387–431.

Johnson, Mark (1987) *The Body in the Mind: The Bodily Basis of Meaning, Imagination, and Reason*. Chicago: University of Chicago Press.

Kluver, Heinrich (1966) *Mescal and Mechanisms of Hallucination*. Chicago: University of Chicago Press.

La Barre, Weston (1969) *The Peyote Cult*. Norman: University of Oklahoma Press.

Leuba, J. H. (1896) A Study in the Psychology of Religious Phenomena. *American Journal of Psychology* 7: 309–85.

Levy, Jerrold, Raymond Neutra, and Dennis Parker (1987) *Hand Trembling, Frenzy Witchcraft, and Moth Madness: A Study of Navajo Seizure Disorder*. Tucson: The University of Arizona Press.

Mayeux, Richard, Jason Brandt, Jeff Rosen, and Frank Benson (1980) Interictal Memory and Language Impairment in Temporal Lobe Epilepsy. *Neurology* 30: 120–5.

Merleau-Ponty, Maurice (1962) Phenomenology of Perception. James Edie, trans., Evanston, IL: Northwestern University Press.

Ortner, Sherry (1974) Is Female to Male as Nature is to Culture? In Michelle Zimbalist Rosaldo and Louise Lamphere, eds., *Woman, Culture, and Society.* Stanford: Stanford University Press.

Persinger, Michael (1983) Religious and Mystical Experiences as Artifacts of Temporal Lobe Function: A General Hypothesis. *Perceptual and Motor Skills* 57: 1255–62.

   (1984a) People Who Report Religious Experiences May Display Enhanced Temporal Lobe Signs. *Perceptual and Motor Skills* 58: 963–75.

   (1984b) Propensity to Report Paranormal Experiences is Correlated with Temporal Lobe Signs. *Perceptual and Motor Skills* 59: 583–6.

Persinger, Michael and P. M. Vaillant (1985) Temporal Lobe Signs and Reports of Subjective Paranormal Experiences in a Normal Population: A Replication. *Perceptual and Motor Skills* 60: 903–9.

Reichard, Gladys (1944) *Prayer: The Compulsive Word.* Seattle: University of Washington Press.

Roberts, J. K. A., M. M. Robertson, and M. R. Trimble (1982) The Lateralising Significance of Hypergraphia in Temporal Lobe Epilepsy. *Journal of Neurology, Neurosurgery, and Psychiatry* 454: 131–8.

Slater, E., A. W. Beard, and E. Glithero (1963) The Schizophrenia-Like Psychoses of Epilepsy. *British Journal of Psychiatry* 109: 95–150.

Tucker, Gary, Trevor Price, Virginia Johnson, and T. McAllister (1986) Phenomenology of Temporal Lobe Dysfunction: A Link to Atypical Psychosis – A Series of Cases. *Journal of Nervous and Mental Disease* 174: 348–56.

Waxman, Stephen G. and Norman Geschwind (1974) Hypergraphia in Temporal Lobe Epilepsy. *Neurology* 24: 629–36.

   (1975) The Interictal Behavior Syndrome of Temporal Lobe Epilepsy. *Archives of General Psychiatry* 32: 1580–6.

Witherspoon, Gary (1977) *Language and Art in the Navajo Universe.* Ann Arbor: University of Michigan Press.

Young, Robert and William Morgan, Sr. (1987) *The Navajo Language: A Grammar and Colloquial Dictionary.* Albuquerque: University of New Mexico Press.

Index